THE CONFLICTED
SUPERPOWER

A NANCY BERNKOPF TUCKER AND WARREN I.
COHEN BOOK ON AMERICAN–EAST ASIAN RELATIONS

NANCY BERNKOPF TUCKER
and WARREN I. COHEN
Books on American-East Asian Relations

Edited by

Thomas J. Christensen

Mark Philip Bradley

Rosemary Foot

Michael J. Green, *By More Than Providence: Grand Strategy and American Power in the Asia Pacific Since 1783*
Jeanne Guillemin, *Hidden Atrocities: Japanese Germ Warfare and American Obstruction of Justice at the Tokyo Trial*

Nancy Bernkopf Tucker was a historian of American diplomacy whose work focused on American–East Asian relations. She published seven books, including the prize-winning *Uncertain Friendships: Taiwan, Hong Kong, and the United States, 1945–1992*. Her articles and essays appeared in countless journals and anthologies, including the *American Historical Review*, *Diplomatic History*, *Foreign Affairs*, and the *Journal of American History*. In addition to teaching at Colgate and Georgetown (where she was the first woman to be awarded tenure in the School of Foreign Service), she served on the China desk of the Department of State and in the American embassy in Beijing. When the Office of the Director of National Intelligence was created, she was chosen to serve as the first Assistant Deputy Director of National Intelligence for Analytic Integrity and Standards and Ombudsman, and she was awarded the National Intelligence Medal of Achievement in 2007. To honor her, in 2012 the Woodrow Wilson International Center for Scholars established an annual Nancy Bernkopf Tucker Memorial Lecture on U.S.–East Asian Relations.

Warren I. Cohen is University Distinguished Professor Emeritus at Michigan State University and the University of Maryland, Baltimore County, and a senior scholar in the Asia Program of the Woodrow Wilson Center. He has written thirteen books and edited eight others. He served as a line officer in the U.S. Pacific Fleet, editor of *Diplomatic History*, president of the Society for Historians of American Foreign Relations, and chairman of the Department of State Advisory Committee on Historical Diplomatic Documentation. In addition to scholarly publications, he has written for the *Atlantic*, the *Baltimore Sun*, the *Christian Science Monitor*, *Dissent*, *Foreign Affairs*, the *International Herald Tribune*, the *Los Angeles Times*, *The Nation*, the *New York Times*, the *Times Literary Supplement*, and the *Washington Post*. He has also been a consultant on Chinese affairs to various government organizations.

ANDREW B. KENNEDY

THE CONFLICTED
SUPERPOWER

America's Collaboration with China
and India in Global Innovation

Columbia University Press / New York

Columbia University Press
Publishers Since 1893
New York   Chichester, West Sussex
cup.columbia.edu
Copyright © 2018 Columbia University Press

Library of Congress Cataloging-in-Publication Data
Names: Kennedy, Andrew Bingham, 1968– author.
Title: The conflicted superpower : America's collaboration with China
and India in global innovation / Andrew Bingham Kennedy.
Description: New York : Columbia University Press, [2018] |
Includes bibliographical references and index.
Identifiers: LCCN 2017048743 (print) | LCCN 2018003615 (ebook) |
ISBN 9780231546201 (electronic) | ISBN 9780231185547 (cloth : acid-free paper)
Subjects: LCSH: United States—Foreign economic relations—21st century. |
Globalization—Political aspects—United States. | India—Foreign economic
relations—21st century. | China—Foreign economic relations—21st century.
Classification: LCC E895 (ebook) | LCC E895 .K46 2018 (print) |
DDC 327.73009/05—dc23
LC record available at https://lccn.loc.gov/2017048743

Columbia University Press books are printed on permanent
and durable acid-free paper.
Printed in the United States of America

Cover design: Diane Luger
Cover image: © ryanking999

For Shameem, Sanya, and Jasper

CONTENTS

CONTENTS

5. THE (MOSTLY) OPEN DOOR:
Global R&D
132

CONCLUSION
155

PREFACE

I F THIS IS a book about innovation, writing it felt like an innovation in my academic career. My first book focused on the foreign policies of China and India, with the United States looming in the background. In this book, the emphasis is reversed. As with many innovations, the roots of this one are apparent in retrospect. American politics has always loomed large in my life, particularly as I grew up surrounded by legislators and lobbyists in the suburbs of Washington, D.C. When I chose to pursue a master's degree at the Fletcher School in my mid-twenties, it was an easy choice to specialize in U.S. foreign policy, even as I became increasingly interested in Asia. In the years that followed, I worked as a consultant in Hong Kong, Beijing, and Washington, scrutinizing U.S. political and economic relations with Asian countries. It was my passion for U.S. foreign policy, particularly toward Asia, that drove me to become an academic in the first place.

This book is about one aspect of U.S. foreign policy: its approach to the globalization of technological innovation. I have focused on this issue not only because it is enormously important, but also because it is so poorly understood. The importance part is clear. For more than seven decades, technological leadership has sustained U.S. economic and military leadership in the international system. Rising powers like China and India, meanwhile, envy the U.S. track record and now seek to become innovation powerhouses in their own right. These quests for innovation, however, are not conducted in isolation. Instead, the United States, China, and India are increasingly intertwined as they pursue technological leadership, owing to cross-border flows of highly skilled people and high-tech investments that are without precedent in history. The policies of the world's innovation leader are particularly important as this process unfolds, but as I will explain in the introduction, they are also particularly puzzling. To my mind, no one has explained why. For all these reasons, this book focuses on the politics behind U.S. collaboration with China and India in global innovation.

I had a great deal of help from others as I researched and wrote this book. First, I want to give special thanks to the many individuals in the United States, China, and India who agreed to be interviewed for this project. Many of these individuals are mentioned in the notes of the chapters that follow, but some wished to remain anonymous or preferred to speak only on a background basis. I am deeply grateful to all of you.

There are many others to thank as well. I am grateful for feedback on draft portions of the manuscript from an array of very gifted and generous scholars: Bjoern Dressel, David Envall, Taylor Fravel, Luke Glanville, Evelyn Goh, Llewelyn Hughes, Devesh Kapur, Scott Kastner, Adrian Kay, Amy King, Bingqin Li, Darren Lim, Andrew MacIntyre, Ann Neville, Margaret Peters, John Ravenhill, Etel Solingen, Helen Taylor, Joanne Wallis, Fiona Yap, Min Ye, and Feng Zhang. My father, Bingham Kennedy, helped make the introduction far more lucid than it would have been otherwise. A number of research assistants have also provided enthusiastic support over the years: Yu-Hua Chen, Jia Guan, Nan Liu, Rongfang Pan, Nimita Pandey, Aditya Parolia, Zhen Qi, Richard Reid, Haiyang Zhang, and Jiayi Zhang. I have also learned much from conversations with Shiro Armstrong, Tai Ming Cheung, Daniel Costa, Christina Davis, Sumit Ganguly, Dan Gold, David Hart, Ron Hira, Scott Kennedy, Robert Keohane, Anupam Khanna, Xielin Liu, Tanvi Madan, Partha Mukhopadhyay, Pratap Bhanu Mehta, Barry Naughton, Neil Ruiz, Michael Teitelbaum, and Patrick Thibodeau. Last, I want to thank the two anonymous reviewers for Columbia University Press for their very constructive comments on the entire manuscript. Any faults that remain are mine alone.

I am thankful for support from a number of institutions. The Crawford School of Public Policy at the Australian National University has been my academic home since 2010, and its interdisciplinary and highly collegial environment has provided an excellent milieu in which to write this book. The Australian Research Council generously funded much of the research through the Discovery Early Career Researcher Award program from 2013 to 2017. Two terrific organizations hosted me as a visiting scholar in 2013: the University of Pennsylvania's Center for the Advanced Study of India and Indiana University's Research Center for Chinese Politics and Business in Beijing. In 2012, I was very fortunate to attend two week-long workshops on China's rise in innovation at the Institute for Global Conflict and Cooperation at the University of California, San Diego.

At Columbia University Press, Stephen Wesley has been a model editor: patient, responsive, and full of good advice. I am honored that the book was selected as a Nancy Bernkopf Tucker and Warren I. Cohen Book on American–East Asian Relations. I am very grateful to the series editors— Mark Bradley, Tom Christensen, and Rosemary Foot—for their decision to include it.

I have dedicated this book to the three wonderful people with whom I share my life: Shameem, Sanya, and Jasper. I hope that one day Sanya and Jasper will appreciate how much they inspire their father—and how profoundly fortunate he was to marry their mother.

THE CONFLICTED
SUPERPOWER

INTRODUCTION

F OR DECADES, SCIENCE-FICTION writers have imagined universal translation devices that render foreign languages instantly intelligible. In 2015, the real world took a step closer to such a technology with the rollout of Skype Translator. Skype was founded in Estonia in 2003, became popular worldwide soon thereafter, and was purchased by Microsoft in 2011. The addition of the translation feature, if still a work in progress, was remarkable. But just as fascinating was how it was invented. Dozens of Microsoft researchers in the United States and China collaborated on the project for years, under a machine-translation effort led by an Indian-born scientist educated in Mumbai and Palo Alto. Fittingly, a technology to facilitate international communication was crafted through cross-border collaboration.[1]

The United States remains the world leader in technological innovation, but as the invention of Skype Translator and many other examples attest, U.S. leadership increasingly involves collaboration with other countries. This development reflects two distinct trends. First, the flow of brainpower—the ultimate ingredient in innovation—from other countries to the United States is nothing short of extraordinary. In 2013, foreign-born individuals accounted for 27 percent of college-educated workers in science and engineering (S&E) occupations in the United States, and 42 percent of all

S&E workers with doctoral degrees.[2] That same year, foreign students on temporary visas received 37 percent of all S&E doctorates granted by U.S. universities.[3] Second, even as the United States attracts high-tech talent, its companies and universities are looking abroad, investing in research and development (R&D) centers overseas and partnering with foreign collaborators. Microsoft is hardly alone in this. By 2014, IBM Research had twelve labs on six continents, and India was home to seventeen of Cisco's top eighty-four engineers.[4] Moreover, these two developments—the migration of brainpower and cross-border R&D—are intertwined. The migration of technically savvy talent, particularly between developing and developed countries, has helped leading firms and universities expand their R&D activities around the world.

This book is about the politics behind these two trends. More specifically, it is about the politics behind the flows of brainpower and R&D between the world's most powerful country—the United States—and its two most prominent emerging powers: China and India. It is often suggested that dominant states and rising powers are doomed to mistrust each other, as the possibility of future conflict casts a shadow on their relations.[5] Although such mistrust exists today, particularly between the United States and China, it has not prevented a remarkable new form of economic exchange from emerging between them. China and India are unrivaled in the number of well-educated workers and students they send abroad, and the United States is the most popular destination for these individuals. U.S. companies, meanwhile, have led the way in setting up R&D centers in China and India, as well as in establishing R&D alliances with Chinese and Indian partners. In important ways, the globalization of innovation is a story of collaboration between the world's most powerful country and its two most important rising powers.

What are the forces behind the growing cooperation in innovation between the United States and these two emerging powers? Whereas commerce has been growing between these countries for decades, innovation is a particularly sensitive arena with important implications for the distribution of wealth and power around the world. One might attribute the globalization of corporate R&D to the diffusion of tougher intellectual property standards internationally, which make it easier for multinationals to protect the fruits of their research abroad. Since the 1980s, in fact, the United States has spearheaded an effort to create, strengthen, and enforce intellectual property standards around the world.[6] In 1994, this movement led to the negotiation of the Agreement on Trade-Related Aspects of Intellectual Property Rights (TRIPS) as part of the World Trade Organization. Yet this is a poor explanation for the growth in U.S. collaboration with China and India, since concerns about intellectual property remain quite pronounced with regard to these countries, particularly China. Indeed, the level of concern that has arisen regarding the protection of intellectual property in recent years makes the growth in U.S. collaboration with these countries all the more puzzling.

Part of the answer, of course, lies in technology itself. The rapidity of international travel, along with revolutionary information and communications technologies, has allowed cross-border movement and communication on an unprecedented scale. Rapid development and rising education levels in China and India have also created huge reservoirs of intellectual talent that have never existed before. Even so, global innovation would not be possible without supportive government policies. These include policies regarding openness to cross-border flows of human capital and transnational R&D activity.

As I will explain, China and India have adopted liberal policies toward global innovation in an effort to develop their economies. With regard to flows of human capital, the Chinese and Indian governments have come to support their professionals and students going abroad, as worries about "brain drain" have receded and enthusiasm for "brain circulation" has grown. In recent years, some have even suggested that a "reverse brain drain" is carrying some Chinese and Indian graduates and professionals back to their home countries, although there is much more evidence of this in China's case.[7] With regard to flows of R&D, both China and India have generally welcomed foreign R&D centers and collaboration with foreign partners in the hope that these actions will have positive spillovers for their own economies. Although the two countries' approaches differ in significant ways, both countries have been broadly open to global innovation.

U.S. policies, in contrast, are more conflicted. The United States places no limit on the number of foreign students who can attend U.S. universities, and it continues to enroll more students from abroad than any other country. The United States is also largely open to overseas R&D, although in some cases it has made efforts to limit the "offshoring" of service jobs abroad. At the same time, the United States is less open to well-educated workers from overseas, and its degree of openness in this regard has varied widely over time. Whereas the annual limit on employment-based (EB) immigrant visas has not changed since 1990, the cap on "specialty workers" through the H-1B visa program (many of whom work for high-tech companies) has risen and fallen dramatically. This has helped create massive backlogs in applications for permanent residency and a dispiriting situation for many talented workers who wish to remain in the country—encouraging some to return home. In some regards, then, the United States has been poorly served by its policies.

U.S. policies are all the more puzzling in the way they defy trends in political relations with China and India. Relations between the United States and China have fluctuated throughout the post–Cold War period; however, particularly since the confrontation over Taiwan in the mid-1990s, worries about military conflict have vexed the relationship. The deterioration in ties from 2003—when then U.S. Secretary of State Colin Powell said relations were "the best they have been since President Nixon's first visit"—to more recent years is palpable.[8] In contrast, the relationship between the United States and India has warmed considerably over the past two decades, and commentators now debate whether the two countries qualify as "allies." Despite these divergent trends, the pattern of U.S. openness effectively privileges China over India. The United States is most open in the arena in which China looms largest: access to higher education. The United States is least open, in contrast, where India predominates: the provision of skilled workers through the H-1B visa program.

In short, the United States is conflicted when it comes to global innovation, and its pattern of openness is puzzling. What are the political forces driving U.S. openness in this realm? Why is the United States more open in some areas than in others, and why has its openness to skilled workers varied so widely over time? And what is the broader significance of these U.S. policies? In the chapters that follow, I will answer each of these questions.

THE ARGUMENT

The policies of interest in this study are hardly unexplored territory for scholars, but the work done so far has not answered the questions I have posed. Recent research has illuminated how developed countries, including the United States, approach high-skill immigration, but these studies have not considered flows of students or explored the globalization of R&D by multinational corporations.[9] There is also a nascent literature on the politics of offshoring, but it has not been concerned with the immigration policies of interest in this study.[10] Scholars have also explored the politics behind how states pursue innovation, but this work has not been concerned with U.S. policies toward immigration or offshoring.[11] In short, we still lack a comprehensive framework to explain U.S. policies toward cross-border flows of brainpower and R&D—the policies that support global innovation.

This book aims to fill that gap. I start by explaining how the globalization of innovation has emerged in recent decades and how China and India have attempted to profit from it. In the remainder of the book, I explain how the United States has chosen to participate in this process. The argument here begins by highlighting the importance of information and communications technology (ICT) in particular, given its status as the world's "leading sector" and its unique role in generating economic and military power today. By leading the way in the ICT revolution, the United States has sustained its primacy in recent decades. Yet it has also become home to an impressive collection of high-tech actors with a strong sense of their own interests. These actors include the world's leading ICT firms and research universities, or what I call the "high-tech community" (HTC).[12] Faced with a globalizing world, these interests have pressed the U.S. government to maintain or develop policies that allow them to take advantage of this development. In immigration, ICT firms and research universities have sought liberal policies that allow them to access talented labor—whether employees or graduate students—from around the world. With regard to R&D, high-tech firms have sought to maintain the latitude to conduct R&D abroad, even if that means offshoring work previously done at home. Simply put, the HTC constitutes a powerful and persistent force for openness in the U.S. approach to global innovation.

The HTC is not omnipotent, however, and it sometimes faces concerted resistance in its quest for openness. In this book, I argue that the type of organized resistance faced by high-tech interests powerfully shapes just how open national policy is. First, there may be no resistance, in which case national policy is generally open. This dynamic illuminates U.S. policy toward the flow of international students, which remains uncapped and continues to increase. Second, the HTC may face resistance from labor. In such cases, national policy is more controversial and openness more qualified, but labor's relative weakness means that constraints on commerce remain limited. This dynamic sheds light on U.S. policy regarding the immigration of skilled workers from the mid-1990s until 2004. U.S. policy toward the offshoring of R&D and other business services can also be understood in these terms. Third, the HTC may face resistance from citizen groups that can mobilize large numbers of voters. In this case, the HTC has greater difficulty achieving its goals, and national policy is likely to be less open than in the first two cases. This dynamic illuminates U.S. policy toward the immigration of skilled workers since 2005.

Overall, this book argues that U.S. openness generally reflects the power and interests of the HTC and that the degree of openness reflects the nature of the resistance the HTC faces from other organized groups.

THE FOCUS

The bulk of this study is concerned with a specific set of U.S. policies that support global innovation. These include policies toward the admission of foreign knowledge workers, the admission of foreign students, and the offshoring of services (particularly R&D). These are policies aimed at the world in general, rather than China or India in particular. Yet because China and India are such prominent partners for the United States, these policies have important implications for U.S. relations with these two countries. In many ways, in fact, U.S. policies toward global innovation are also U.S. policies toward China and India. A U.S. policy restricting the inflow of foreign S&E graduate students would affect China more than any other country, and a U.S. policy altering the inflow of knowledge workers would affect India more than any other country.

As mentioned, U.S. policies toward global innovation are quite puzzling. It is partly for this reason that most of this book focuses on the United States. Yet there are also several other reasons to focus on U.S. policymaking. First, although the globalization of innovation is a worldwide phenomenon, the United States represents a critical focal point. Since the mid-twentieth century, the United States has been the world leader in innovation, and in recent decades, it has underscored this status by leading the world into the ICT era. The United States is also deeply involved in the globalization of innovation, in terms of both the flow of brainpower and R&D activity. But despite the profound importance of the United States in global innovation, we have little insight into how the country approaches this phenomenon.

Second, comparing different arenas of U.S. policymaking makes sense from the perspective of research design. The HTC faces varying types of opposition in each policy domain considered in this study, allowing for ample variation in the independent variable of interest. Yet because each of these domains is situated within the American context, they are broadly comparable: These are policies developed within a single political system and during a single time period. To that extent, this study approximates a "controlled comparison."[13] To be sure, there are some inherent difficulties in comparing policies toward outward foreign investment with those toward legal immigration, particularly because the former are inherently more difficult to regulate than the latter. As will be discussed, however, these difficulties are not insurmountable. Moreover, in one case in particular—U.S. policy toward the flow of well-educated workers—the HTC has faced a varying type of opposition over time, which allows us to make a "before-and-after" comparison within a single policy domain. The possibility remains, of course, that other variables may explain the outcomes observed in this study, a point to which we will return in chapter 2.

Third, many sources of information can be used to investigate U.S. policymaking toward global innovation. The policies examined in this study have all been sources of controversy in recent decades. The admission of well-educated workers to the United States has featured in recurrent legislative fights; the inflow of foreign students was scrutinized after the September 11, 2001, terrorist attacks; and the phenomenon of offshoring has stoked anxieties among white-collar workers since the early 2000s. This study takes advantage of the extensive media coverage of the lobbying and legislation involved in each of these cases. In addition, because this study focuses on relatively recent events, it has been possible to interview many key participants in the policymaking process. Interviews for this study were conducted with seventy-two individuals between 2013 and 2017 in the United States, China, and India. These individuals include current and former government officials from these countries, as well as representatives of high-tech firms, business associations, organized labor, citizen groups, and other relevant organizations.[14] Lastly, this study draws on quantitative evidence from a range of public and private databases, including those maintained by the U.S. Department of Commerce, the U.S. Department of State, the National Science Foundation, the Center for Responsive Politics, and Thomson Reuters.

WHY IT MATTERS

The paucity of research on the politics of global innovation is understandable, given how new the phenomenon is. Even so, it is an important gap in our understanding of international politics for a number of reasons. Most broadly, global innovation matters because technological creativity is a basic driver of human development. New technologies change how people communicate, travel, work, and do virtually everything else in life. For all these reasons, technological innovation represents a key source of economic growth. Indeed, economists have focused on innovation as a crucial driver of national economic development, particularly in recent decades.[15] Whereas increasing inputs of capital and labor face diminishing returns, innovation increases the efficiency with which such inputs are used and thus represents a sustainable source of long-term growth. Globalization poses new opportunities for innovation, particularly by enabling new sources of human capital and creating new locales in which to do R&D. Yet it also creates new challenges, including political challenges as innovation comes to rely on economic openness in a variety of countries around the world.

Global innovation also has important implications for international politics. First, global innovation knits economies together more tightly and increases the degree of interdependence among them. In the information age, the generation of new knowledge—and intellectual property—is the most valuable form of economic activity. To the extent that the United States is collaborating with China and India in the creation of valuable knowledge, its economic relationships with these countries are becoming costlier to disrupt. This increasing degree of collaboration does not mean that these relationships will become increasingly free of political tension, however: Recent scholarship has shown that the strength of commercial and political ties can vary independently of each other.[16] Yet deeper economic interdependence may well decrease the chance of armed conflict, or at least some forms of armed conflict, between the United States and other powers.[17] Indeed, research on the globalization of production—including innovation—maintains that it has a pacifying effect among the most powerful states.[18] Other scholars have argued that it is expectations of future commerce that generate peace between states.[19] Compared with other forms of economic cooperation, collaboration in innovation would seem particularly likely to foster beliefs that commerce will continue, since investments in innovation are often long term in nature.

The globalization of innovation also has implications for the distribution of power. International relations scholars have long recognized that prowess in technological innovation can underpin national power. Robert Gilpin, for example, has emphasized that major advances in technology allow new nations to rise to political preeminence, though over time, technological know-how and "inventiveness" diffuse to other countries.[20] "Long-cycle" theorists, in turn, have argued that new countries become dominant because they develop innovations in new commercial and industrial spheres, or "leading sectors," which undergird the dominant state's economic vitality and military power.[21] To the extent that U.S. openness to global innovation accelerates the diffusion of "inventiveness," as Gilpin put it, it should accentuate the rise of emerging powers like China and India. This is not hard to imagine: The intimate collaboration between U.S. firms and universities on the one hand and Chinese and Indian workers and students on the other creates possibilities that have never existed before, particularly when the latter return home or gain employment with firms from their own countries. Moreover, although the U.S. government continues to control the spread of sensitive military technologies, it has become increasingly difficult to differentiate between "civilian" and "military," and many technologies that are useful in both spheres are uncontrolled.[22]

Nonetheless, the globalization of innovation also has the potential to reinforce U.S. economic leadership. In particular, it can be understood as a means through which the United States attracts human capital from around the world, including from China and India. U.S. universities prosper from the flow of talented students to their campuses, and many of these students remain in the United States after graduation. U.S. companies have the opportunity to hire the most talented workers—either within the United States or overseas when they invest in R&D abroad. It is thus possible that global innovation will work to the advantage of the United States in a number of ways, rather than to its detriment.

PLAN OF THE BOOK

This introduction is followed by five chapters and a conclusion. Chapter 1 sets the stage by describing global innovation in more detail. The chapter begins by explaining the roles played by cross-border flows of human capital and R&D activity in global innovation, while also considering how far each of these trends has progressed. The chapter then describes how China and India have come to embrace global innovation over the past few decades. The discussion reveals that both countries have approached global innovation as an opportunity to further their own development, but it also highlights the distinct ways each country has done so.

Chapter 2 develops a theory to explain why the world's dominant state would adopt varying levels of openness in its policies toward global innovation. The chapter begins by explaining why the world's innovation leader and its dominant state have traditionally been one and the same. It emphasizes the importance of leading sectors and notes how the United States predominates in today's leading sector: ICT. The chapter then introduces the concept of the high-tech community and explains how this collection of interests has become particularly prominent in the ICT revolution of the past several decades. The chapter goes on to develop a theory to explain variation in the dominant state's degree of openness to global innovation, focusing on the preferences of the HTC and the relative power of its opponents, while also outlining some alternative explanations for its policies. The chapter concludes by explaining how the various explanations will be tested in the remainder of the book.

Chapters 3, 4, and 5 test the theory developed in chapter 2 through a series of case studies focused on U.S. policies from the mid-1990s through 2016. Chapter 3 focuses on U.S. openness toward foreign knowledge workers. The chapter begins with an overview of U.S. policy, particularly toward EB visas and H-1B visas. The discussion notes how the annual cap on H-1B visas was raised in 1998, 2000, and 2004, and how subsequent attempts to raise the cap have fallen short. To explain this variation, the chapter delves into the political battles that have driven U.S. policies toward skilled immigration since the mid-1990s. Between 1998 and 2004, for example, the HTC faced resistance mainly from organized labor, a lopsided contest in which the HTC largely prevailed. Since 2005, however, the HTC has also faced opposition from large citizen groups, as skilled immigration has become embroiled with broader immigration legislation. Because these groups are more formidable opponents, the HTC has failed to increase skilled immigration for more than a decade.

Chapter 4 focuses on U.S. openness to foreign students. The chapter begins by explaining that U.S. policy is far more open in this regard than with regard to foreign knowledge workers: Unlike EB visas and H-1B visas, student visas are not subject to an annual cap. The discussion here also notes the prominent role that China and India play in supplying students—especially graduate students in science and engineering fields—to American universities through the F-1 visa program. The chapter then explores the politics behind U.S. policy, focusing on the efforts of HTC members (particularly universities) to maintain an open policy toward foreign students. The analysis focuses on U.S. policy in the wake of the September 11, 2001, terrorist attacks, when security concerns threatened to impede the inflow of foreign students. The chapter shows how the response of the HTC, combined with the near total absence of organized opposition, played an important role in sustaining U.S. openness to foreign students.

Chapter 5 explores U.S. openness to global R&D. It focuses in particular on policy toward outward foreign investment, since it is in this realm that global R&D has faced its most serious challenge. The chapter begins by explaining how the United States has traditionally been quite open to outward investment, but it also notes a range of concerns that have arisen in recent years, particularly regarding the impact of offshoring on U.S. employment and technological leadership. It then explores the politics behind U.S. policy. The analysis shows that although American lawmakers at both the federal and state levels have had opportunities to limit offshoring, they have taken relatively few of them. To explain this outcome, the chapter shows how HTC members (particularly high-tech firms) have worked with other business groups to block anti-offshoring legislation. Because the HTC has faced resistance only from organized labor in this case, it has been largely (if not totally) successful.

The conclusion puts the book's findings in a broader context. First, it sums up the findings and assesses them in terms of the theory outlined in chapter 2. It then considers the implications for international relations and, in particular, the study of global innovation going forward. Next, it explores whether the interest-group battles examined in the book represent a form of political decay that is accelerating American decline vis-à-vis China and India, or whether other interpretations are possible. The book concludes by considering the implications of the outcome of the 2016 U.S. federal elections and how U.S. policies may evolve in the future.

1

THE RISE OF GLOBAL INNOVATION

I F TECHNOLOGICAL CREATIVITY often concentrates in particular countries, the processes that undergird innovation increasingly transcend lines drawn on the map. Some of the world's most innovative organizations, particularly U.S. companies and universities, have played a pioneering role in this transformation through overseas investments and alliances, enthusiastic recruitment of foreign workers, and large-scale admissions of foreign students. The world's two most prominent rising powers, meanwhile, have sought to harness these trends as they seek to become innovation leaders in their own right. This does not mean that China and India will succeed: Global innovation poses significant challenges for both of these countries, and both must continue to reform domestically.[1] It does mean, however, that the United States has helped to create an international environment with no parallel in history.

In this chapter, I explore the contours of global innovation as it has emerged in recent decades. In the first section, I begin by explaining how I use the terms *innovation* and *global innovation*. I then highlight two distinct aspects of global innovation: the globalization of brainpower and the globalization of research and development (R&D) activity. I pay particular attention to the leading role played by the United States and to information and communications technology (ICT) industries in both regards.

In the latter part of the chapter, I explore how China and India have opened themselves to global innovation and, in doing so, have tried to take advantage of it. I also examine the distinct ways in which China and India have approached global innovation, particularly in terms of greater activism on the part of the Chinese state and India's greater reliance on diaspora labor.

GLOBAL INNOVATION: AN OVERVIEW

Innovation can be defined in many ways. Following Joseph Schumpeter, scholars often distinguish between a number of phases in the innovation process: the initial invention of a novel product or process, efforts to commercialize the invention or otherwise put it into practice, the diffusion of the invention, and the imitation of the invention by others.[2] In this book, I focus on transnational processes that support firms' and universities' efforts to engage in the first of these tasks: the creation of a product or process that is "new to the world."[3] Such new-to-the-world technologies can take a variety of forms, of course, and it is common to distinguish between "radical" and "incremental" innovations.[4] Radical innovations are what we often imagine when the word *innovation* is used—the invention of a wholly new technology (such as the airplane) or way of organizing production (such as the assembly line). Incremental innovations are less dramatic, but they can still be terrifically important. Incremental innovations are often needed to make a new technology useful, and the cumulative effect of a series of incremental innovations over time can be profound.[5]

What are the transnational processes that support innovation, as I define it here? In the past, scholars have frequently focused on the "global innovation networks" of firms.[6] These may be intra-firm networks, in which multinationals assign certain R&D tasks to offshore affiliates, or they may be inter-firm networks in which multiple firms collaborate. In this book, I take a broader approach. Specifically, I am interested not only in transnational R&D activity, but also in cross-border movements of high-tech labor.[7] As I will explain, the latter is both important and extensive in scale—as well as intimately intertwined with the globalization of R&D activity. When I use the term *global innovation*, therefore, I am referring to the transnational activity of multiple inputs into the innovation process, involving both capital and labor.[8]

THE GLOBALIZATION OF BRAINPOWER

In a fundamental sense, people are at the heart of innovation. Yet innovation requires the right sort of people: intelligent, creative, educated, and diligent. Or to put it in more social scientific terms, innovation requires people with the right sort of human capital: labor with the skills, education, and personality attributes to push back the technological frontier. Scholars of innovation have recognized this fact for many years. In fact, a classic paper by Paul Romer posited that the generation of new knowledge was a function of only two inputs: the existing stock of knowledge and human capital.[9] Of course, human capital can refer to a range of attributes, from the percentage of a population that is literate to the fraction with doctoral degrees in scientific disciplines. Because the focus of this study is on new-to-the-world innovation, I am interested here in the more highly educated end of the spectrum. It is this sort of human capital that allows states to exercise technological leadership.

It is useful to distinguish between two types of highly educated labor. The first consists of professionals who have completed their education and are hired by an organization to conduct R&D. In some industries, such as biotechnology, these are frequently individuals with graduate degrees in their field of expertise. In other industries, such as information technology, undergraduate degrees can provide the requisite expertise. In the United States, the private sector is the largest employer of such labor in the economy. In 2013, for-profit businesses employed 62 percent of the 5.7 million scientists and engineers employed in science and engineering (S&E) occupations in the country. Higher education employed an additional 18 percent, the public sector employed 12 percent, and nonprofit organizations employed 5 percent.[10] The most likely to work for businesses were engineers (76 percent) and computer scientists (73 percent).[11]

The second type of highly educated labor consists of academic labor, particularly graduate students and postdoctoral researchers. Whereas academic research in S&E fields is led by senior scientists, the bulk of the labor pool consists of graduate students and postdoctoral researchers. In 2013, for example, U.S. academia employed 144,400 full-time S&E faculty who listed research as their primary or secondary activity.[12] In contrast, there were more than 457,000 full-time graduate students and roughly 43,000 postdoctoral researchers in S&E disciplines in the United States that year.[13] The collaborative nature of scientific research means that these graduate students and postdocs are frequently crucial to the execution of important laboratory research. In fact, income from research assistant work was the primary source of income for nearly 115,000 S&E graduate students in the United States in 2013.[14] Academic research, meanwhile, can play an important role in the broader national innovation effort.[15] To give one famous example, in the late 1950s, Leonard Kleinrock's dissertation at the Massachusetts Institute of Technology (MIT) addressed the challenge of allowing computers to communicate with each other, and in 1969, Kleinrock's laboratory at the University of California, Los Angeles (UCLA), sent the first message over the Advanced Research Projects Agency Network (ARPANet), the precursor to the Internet. Top universities may also help create "regional clusters" that become key sites for innovative firms to locate. In the classic examples, Stanford University and the University of California, Berkeley have helped Silicon Valley thrive, whereas MIT and Harvard University support the Route 128 cluster outside Boston.

Whether well-educated workers or students, cross-border flows of brainpower have a long history. In the mass migrations of the nineteenth and early twentieth centuries, most immigrants were unskilled, but there were exceptions.[16] In the late nineteenth century, British firms often hired German chemists, and some of the most successful entrepreneurs in the British chemical industry were of German origin.[17] Later, in the 1930s and 1940s, Jewish scientists fled Nazi Germany for the United States—and were followed by a clandestine flow of Nazi scientists after World War II— promoting U.S. scientific and technological leadership in the post-war era.[18] The migration of brainpower became more pronounced in the 1960s, particularly with the easing of racial aspects in the immigration laws of the United States, the United Kingdom, Canada, and Australia. Large flows of human capital from developing to developed countries ensued, and "brain drain" emerged as a concern.[19]

Since 1990, cross-border flows of human capital have reached unprecedented heights. By one estimate, the stock of skilled immigrants—that is, those with some postsecondary education—in member countries of the Organisation for Economic Co-operation and Development (OECD) increased by 64 percent between 1990 and 2000, with the number from developing countries increasing by 93 percent.[20] Between 2000/01 and 2010/11, meanwhile, the number of immigrants with postsecondary educations in OECD countries rose by 70 percent to 35 million.[21] The top sources of these immigrants were India (2.2 million), the Philippines (1.5 million), and China (1.5 million). The United States, with about a third of the total, was home to the largest share.[22]

These flows of human capital are no longer simply a unidirectional brain drain from developing to developed countries. Well-educated migrants now frequently visit or return to their home countries, and in some cases the brain drain turns into "brain circulation."[23] Even so, many of these well-educated individuals stay abroad and become an important part of the S&E labor force in their new OECD homes. This is true even in a country as large as the United States. By 2013, foreign-born individuals accounted for 27 percent of all college-educated workers employed in S&E occupations in the United States, 34 percent of all S&E workers with master's degrees, and 42 percent of all S&E workers with doctoral degrees (figure 1.1). Foreign labor is particularly prominent in the ICT industry. In 2013, foreign-born workers made up more than 50 percent of those employed in S&E occupations in the United States with a doctoral degree in computer or mathematical sciences.[24] Silicon Valley is a particular magnet for such workers. In 2015, foreign-born workers comprised a remarkable 67 percent of employees aged 25 to 44 years in computer and mathematical occupations in Silicon Valley, according to one estimate.[25]

The number of international students has soared as well. In the fifteen years from 1975 to 1990, the number of foreign students worldwide increased

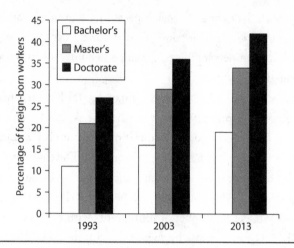

FIGURE 1.1 Percentage of foreign-born workers in the United States S&E occupations by education level: 1993, 2003, 2013. *Source*: National Science Foundation, *Science and Engineering Indicators 2016* (Arlington, VA: National Science Foundation, 2016), chapter 3, 101.

from 800,000 to 1.3 million.[26] Over the following fifteen years, however, the total more than doubled to 3 million. And from 2005 to 2012, the number jumped to 4.5 million. The rising number of students from Asia accounts for much of this increase: 53 percent of all international students in 2012 were from Asian countries.[27] China was easily the leading source, accounting for 22 percent of all foreign students enrolled in schools in OECD countries. India was the second largest source, accounting for 6 percent of all foreign students studying in OECD countries. The most popular destination for students from both countries was the United States.[28]

The number of foreign graduate students in S&E fields has also increased dramatically. This is quite evident in the top three destination countries for international students as of 2010: the United States, the United Kingdom, and Australia.[29] In Australia, the number of foreign students in graduate research programs nearly quintupled between 2002 and 2014. By 2014, foreign students made up more than 38 percent of graduate research students in the natural and physical sciences—and more than 50 percent in engineering and information technology.[30] In the United Kingdom, meanwhile, the proportion of foreign students in graduate S&E programs rose from 29 percent in the mid-1990s to 48 percent by 2013/14.[31]

The growing importance of foreign S&E students is arguably most striking in the United States, however, given the size of its higher education sector. Foreign students on temporary visas received 17 percent of all doctoral degrees awarded in S&E fields by U.S. universities in 1977, but by 2013, that figure was 37 percent.[32] Once again, China and India are the leading countries of origin.[33] The share of foreign-born postdocs, meanwhile, increased from 18 percent in 1983 to 48 percent in S&E and health fields in 2013.[34] Foreign doctoral graduates on temporary visas were particularly prominent in fields relevant to the ICT industry: In computer science, for example, their share was 53 percent.[35] These graduates, in turn, have become important sources of academic expertise in the country. In 2013, foreign-born scholars with U.S. doctorates comprised 50 percent of all those employed in academic computer science in the United States, including 52 percent of full-time faculty in the field.[36] Foreign graduates with degrees in S&E fields are also employed more widely throughout the U.S. economy, of course. In 2013, 2.4 million of the 4.6 million college-educated, foreign-born scientists and engineers employed in the United States received their first bachelor's degree from a U.S. university. Of the 2.2 million who received their first degree elsewhere, nearly 700,000 received a graduate degree from a U.S. university.[37]

Well-educated labor is a critical ingredient in innovation, and the supply of such labor has never been more mobile. Whether one considers professional workers for corporate R&D or candidates for graduate work in leading research universities, a much larger part of the relevant labor pool is being drawn from abroad than ever before.

THE GLOBALIZATION OF R&D

R&D has emerged as central to innovation in the modern era. Whereas in the nineteenth century and earlier, individual geniuses such as Thomas Edison played key roles in invention, by the early twentieth century, industrial research efforts had become the key locus of advances in the chemical and electrical industries and were becoming important more widely.[38] The simplicity of the label "R&D" is deceiving, because it masks a wide variety of inventive activity. Indeed, defining which activities qualify as R&D and which do not can be a daunting endeavor. Citing widespread international practice, the National Science Foundation in the United States distinguishes three components of R&D:

- Basic research: The objective of basic research is to gain more comprehensive knowledge or understanding of the subject under study without specific applications in mind. Although basic research may not have specific applications as its goal, it can be directed in fields of present or potential interest. This is often the case with basic research performed by industry or mission-driven federal agencies.
- Applied research: The objective of applied research is to gain knowledge or understanding to meet a specific, recognized need. In industry, applied research includes investigations to discover new scientific knowledge that has specific commercial objectives with respect to products, processes, or services.
- Development: Development is the systematic use of the knowledge or understanding gained from research directed toward the production of useful materials, devices, systems, or methods, including the design and development of prototypes and processes.[39]

Like the supply of human capital, the performance of R&D is globalizing. This development has been facilitated by advances in communications and transportation technology over the past several decades, as well as the increasingly transnational nature of modern corporations. And like flows of human capital, global R&D takes different forms. These include investments in foreign R&D centers by multinational firms and transnational R&D collaborations, which involve a wide variety of organizations.[40] Let us consider each of these in turn.

Overseas R&D investments have become increasingly common as leading multinationals establish research centers in multiple locations across the globe.

This development has sparked concerns that the innovative capacity of developed countries is "hollowing out" as multinationals take advantage of less expensive, but well-educated, technical workers in the developing world.[41] The reality is not so dramatic. Indeed, for the United States, it is clear that this aspect of global innovation has not developed as quickly as have cross-border flows of brainpower. Between 2000 and 2010, for example, the share of U.S. corporate R&D performed within the United States fell from 88 percent to 84 percent.[42] The very gradual nature of the shift reflects a number of constraints: the embeddedness of multinationals in their home countries, the need for internal cohesion within multinational firms, the challenges of managing global R&D networks, and the local infrastructure and intellectual property protection that R&D requires.[43]

Nonetheless, it is clear that multinationals are grappling with these challenges and investing greater amounts in foreign R&D centers than ever before.[44] In 2013, foreign companies—mostly from other developed countries—spent $54 billion on R&D in the United States.[45] That same year, U.S. companies spent $49 billion on R&D abroad.[46] A majority of that spending (61 percent) took place in Europe, particularly Germany and the United Kingdom (table 1.1). The amount spent in developing countries is clearly growing, however. U.S. companies spent negligible amounts on R&D in China and India for much of the 1990s, but by 2013, the two countries accounted for nearly 10 percent of U.S. R&D spending abroad.[47]

U.S. ICT firms have been leaders in global R&D. IBM established its first research laboratory outside the United States in Switzerland in 1956; others followed in Israel (1972) and Japan (1982). Intel established its first overseas R&D center in Israel in 1974. Microsoft Research Asia was established in Beijing in 1998 and has become the company's second-largest research organization. Taken together, industrial groupings dominated by ICT firms accounted for a substantial share of all U.S. corporate R&D abroad as of 2013. These included computers and electronics products ($7.5 billion), computer systems design ($2.4 billion), and information ($4 billion).[48]

Traditionally, an important distinction has been made between two different types of overseas corporate R&D.[49] The first is asset-exploiting

TABLE 1.1 TOP COUNTRIES FOR OVERSEAS U.S. CORPORATE R&D SPENDING IN 2013

COUNTRY	AMOUNT ($ BILLION)
Germany	8.2
United Kingdom	5.4
Switzerland	3.7
Canada	3.2
Belgium	2.6
India	2.6
France	2.4
Japan	2.4
China	2.2
Israel	2.1

Source: Bureau of Economic Analysis, "U.S. Direct Investment Abroad, All Majority-Owned Foreign Affiliates, Research and Development Expenditures for 2013," *International Data: Direct Investment and Multinational Enterprises*, 2016, www.bea.gov/iTable/index_MNC.cfm.

(or home-base exploiting) R&D. The focus here is on development—but a limited form of development in which existing products are adapted to foreign markets. Such adaptation may be necessary to make a product more competitive or appropriate in the new environment. The second is asset-augmenting (or home-base augmenting) R&D. In this case, the focus is on tapping into local knowledge or competencies that may be globally relevant. There is the possibility of actual research and more sophisticated development here, since the purpose is to improve existing assets or acquire or create new ones.[50]

A second form of global R&D involves the pooling of resources by organizations in different countries to engage in cross-border collaboration. One prominent form of such collaboration involves R&D alliances in which firms, universities, and even governments engage in joint research projects with partners in other countries. R&D alliances were extremely rare in the 1970s, but they became much more common in the 1980s and then exploded in the early 1990s, apparently as a response to the "technological shock" posed by the Internet's emergence.[51] R&D alliances are useful to firms in particular as means to access complementary expertise, to share risk and cost, and to cope with accelerating technological change.[52]

R&D alliance activity has globalized more rapidly than corporate R&D spending has. The Thomson Reuters SDC Platinum database, the most comprehensive repository of data on such alliances, reports that 12,686 R&D alliances (including joint ventures) were created worldwide between 1990 and 2014, 6,448 of which were international in nature (i.e., they involved participants from more than one country). The United States has been very active in this regard, in terms of both total alliance activity and international alliance activity (table 1.2). Traditionally, Western European countries and Japan have been the major sources of foreign partners for the United States in R&D alliances. In recent years, however, China and India have provided a larger share of U.S. partners than in the past, with China's share nearly equaling that of Japan from 2010 to 2014.[53]

There are also other forms of transnational R&D collaboration. Cross-border venture capital (VC) investments have become quite common, whether by VC firms or by corporate VC arms. While European VC firms have long engaged in cross-border investments, this century has seen U.S. VCs become much more active in this regard. A 2011 survey of U.S. VC firms, for example, revealed that 49 percent had already invested overseas at that point. It also found that 72 percent of those surveyed planned to increase or maintain their overseas investments in the future.[54] There is also increasing foreign VC investment in the United States. In the first

TABLE 1.2 U.S. R&D ALLIANCES

TIME PERIOD	TOTAL	INTERNATIONAL	WESTERN EUROPE	JAPAN	CHINA	INDIA
1990–2014	9,894	4,741	2,184	1,272	177	89
2010–2014	376	210	107	24	23	9

Source: Thomson Reuters SDC Platinum database. Data extracted April 2015.

quarter of 2016, for example, 27 percent of all VCs investing in the United States were headquartered abroad.[55] The leading sources were the United Kingdom, China, and Israel. Worldwide, VC investments remain closely associated with the ICT industry. In 2015 and the first quarter of 2016, for example, between 76 percent and 79 percent of all VC-backed deals globally were in ICT.[56]

There are still more forms of collaboration on the academic side. Whereas R&D alliances tend to be relatively focused, some universities have established wider-ranging research partnerships. In 2007, for example, MIT and Singapore's National Research Foundation established the Singapore–MIT Alliance for Research and Technology (SMART), MIT's first research center outside the United States.[57] More generally, one database compiled by two U.S. higher education associations identified 369 international research partnerships concluded by their members as of 2009.[58] Such partnerships help researchers forge long-term collaborations with foreign colleagues and sometimes to access foreign research funding.[59] There are also myriad small-scale collaborations among individual laboratories in different countries, as the data on international co-authorship make clear. Worldwide, the fraction of published S&E articles written by authors from more than one country jumped from 13 percent in 2000 to 19 percent in 2013. Once again, the United States has helped lead the way: The share of all U.S. S&E articles that included a foreign co-author jumped from 19 percent to 33 percent from 2000 to 2013.[60]

In sum, R&D activity is globalizing in a number of ways, though not always as rapidly as flows of brainpower have in recent decades. These two processes, moreover, are intimately intertwined. In China and India, for example, U.S. ICT firms frequently rely on Chinese or Indian staff with experience in the United States to play important roles in their local R&D.[61] To cite one example, the founding director of Microsoft Research Asia, Kai-Fu Lee, was born in Taiwan but earned B.S. and Ph.D. degrees in computer science from Columbia University and Carnegie Mellon University, respectively, and worked in Apple's R&D department for six years before moving to Microsoft.[62] Lee later became the first president of Google China. Similarly, the first director of the IBM India Research Lab in Bangalore was Guruduth Banavar. Banavar studied as an undergraduate at Bangalore University, but he received his Ph.D. in computer science from the University of Utah, and he worked for IBM Research in New York for ten years before taking the Bangalore job.[63] There are more prominent examples as well. In 2004, Yahoo! India hired Prasad Ram as its chief technical officer. Ram received his undergraduate degree from the Indian Institute of Technology, Bombay, but he earned his Ph.D. in computer science from UCLA, after which he worked in the United States as a research scientist for Xerox for six years. Following his time at Yahoo! India, Ram went on to become the head of R&D for Google India from 2006 to 2011.[64]

Migration and global R&D are also intertwined on the academic side. This has become apparent in the relationship between the United States and China. The United States grants more student and academic visitor visas to individuals from China than from any other country.[65] Although many of these individuals stay in the United States, an increasing number are returning to China. In fact, the U.S. J-1 visa, which was granted to 38,928 visitors from China in 2015, frequently requires recipients to return to their home country for at least two years at the end of their program. This returning flow of Chinese scientists, in turn, has helped fuel the rapidly growing collaboration between the U.S. and Chinese scientific establishments. In 2013, Chinese scientists and engineers published nearly 31,000 articles with U.S. collaborators—up from fewer than 2,500 in 1999.[66] As a result, China has emerged as the leading source of scientific co-authors for the United States: In 2013, the number of articles co-authored by Chinese scientists was one and a half times that of those co-authored by British scientists, triple that of those co-authored by Japanese scientists, and six times that of those co-authored by Indian scientists.

The migration of human capital and the globalization of R&D are interconnected processes that together have unleashed a new era of global innovation. The United States, in turn, has played a leading role in driving both processes forward. Yet the world's two most populous countries are also playing key roles in global innovation. China and India have become the preeminent exporters of brainpower to the rest of the world, and they have also begun to play more important roles in global R&D.

CHINA AND INDIA IN GLOBAL INNOVATION

Chinese and Indian leaders have long aspired to technological modernity. In the early years of the Cold War, Mao Zedong and Jawaharlal Nehru sought to transform their countries into advanced economic powers, particularly (if not only) through investments in nuclear technologies.[67] As both countries have reformed and opened their economies in recent decades, they have become increasingly involved in the globalization of innovation, particularly in ICT. In doing so, Chinese and Indian leaders have been well aware of the risks involved. Talented students and workers may not return home—the dreaded brain drain—widening rather than closing the gap with developed countries. Foreign firms setting up R&D centers may monopolize the best domestic talent and have little connection with the local economy.[68] Despite these dangers, both China and India have come to see global innovation as an opportunity. Both countries have treated the outflow of talented students and labor as a chance to boost the human capital of their own economies and participate in the global economy. Both have also come to embrace global R&D, hoping there will be positive spillovers for their own innovation capacities. Although there are significant differences in their approaches, both countries have embraced the globalization of innovation as a chance not to be missed.

CHINA

China has opened to global innovation with no little ambition. In 2006, the government's National Medium- and Long-Term Program for Science and Technology Development (2006–2020) was launched to rapidly advance "indigenous innovation" (*zizhu chuangxin*) and to make China a world-class "science and technology power" (*keji qiangguo*).[69] Although much attention has been devoted to the role of industrial espionage in China's pursuit of these goals, such activity is but one element in China's wider efforts to emerge as an innovation leader.[70] Since the 1970s, these efforts have included participating in international flows of human capital with Western countries. More recently, they have included openness to global R&D.

Flows of Brainpower

Sending students to Western countries has been part of China's plans for economic reform and opening up from the beginning. In the early 1970s, China began sending small numbers of students to Western countries, including Australia, Canada, France, and the United Kingdom.[71] With the onset of economic reform the late 1970s, Deng Xiaoping called for more students to go abroad, particularly to learn more about advanced technologies. Between 1977 and 1983, as a result, 26,000 officially sponsored students and visiting scholars went abroad from China, along with another 7,000 self-financed students. This was double the number of students sent abroad between 1950 and 1977, and more than half of these students went to the United States.[72]

However, Chinese leaders began to have second thoughts about sending so many students abroad as the 1980s progressed. First, there was growing anxiety that Chinese students were not returning after completing their studies. In response to this concern, new policies introduced in 1987 and the first half of 1988 began to restrict opportunities for overseas study and the duration of such study. As close observers at the time put it, these policies reflected "a degree of panic over the potential brain drain."[73] Second, leaders were concerned that returning students were spreading Western ideas that would undermine China's political system. Following student protests in late 1986, in fact, the Chinese Communist Party (CCP) leadership launched an "anti-bourgeois liberalization" campaign aimed at the liberal ideas associated with returning students. Last, and relatedly, leaders were particularly con-cerned about the number of students going to the United States. In 1987, Deng Xiaoping criticized the number of students choosing to study in the United States, prompting the State Education Commission (SEDC) to propose cutting the U.S. share of outgoing Chinese students from 68 percent to 20 percent. The SEDC also proposed increasing pressure on students already in the United States to return to China.[74]

The next few years would see dramatic shifts in China's perspective. In mid-1988, Chinese officials and the state media began to display more confidence that China could manage the challenges posed by overseas study.[75] As one Chinese journalist stated in late 1988, "The key to solving this problem lies in developing the economy, not in changing the overseas students policy."[76] This more confident attitude was short-lived, however. In June 1989, the events surrounding the Tiananmen Square massacre empowered conservative leaders and prompted a rethink of overseas study. In August 1989, an internal circulation report concluded that the United States had been actively working with some Chinese students to sow discord within China.[77] This U.S. strategy, it was charged, was in line with Washington's longstanding desire to promote "peaceful evolution" in China and undermine the country's socialist system. In March 1990, an SEDC report—reportedly approved by Premier Li Peng—made reference to the number of students still in the United States and stated that the U.S. government "holds the exchange students and scholars as hostages."[78]

Deng Xiaoping's efforts to reinvigorate reform in China in the early 1990s would lead to a new approach. Deng was apparently dismayed at China's continuing loss of talent to the United States in the early 1990s, and he determined to improve the climate within China for returnees. During his historic "southern tour" in early 1992, he stated,

> We hope all those studying abroad will come back. All overseas students may return and enjoy proper arrangements, regardless of their previous political attitude. This policy toward them must remain unchanged. They should be told that it's better to return home if they want to make their contributions.[79]

Deng's forgiving tone was controversial among the Chinese leadership at the time.[80] Even so, his line prevailed. In November 1993, the CCP officially adopted a new slogan at the Third Plenum of the Fourteenth Party Congress: "Support overseas study, encourage people to return, and give people the freedom to come and go" (*Zhichi liuxue, guli huiguo, laiqu ziyou*).[81] Notably, the CCP also endorsed "taking a variety of measures to encourage overseas talent to serve the motherland," reflecting a growing recognition that not all students would come back to China. Whereas in 1978 Deng had hoped that perhaps 90 percent of China's overseas students would return, in the early 1990s, he suggested that perhaps only half of those abroad might do so.[82]

In the late 1990s, President Jiang Zemin deepened China's commitment to this approach. Many Chinese students continued to remain overseas throughout the 1990s, and Jiang was well aware of it. In 2000, the Chinese president stated in an interview with the editor of *Science* magazine that only 110,000 of 320,000 Chinese who had gone abroad to study since 1978 had returned to China. The challenge for China, in Jiang's view, was not openness to overseas study but conditions within China itself. "We need to further develop the Chinese economy and create better conditions to attract more people back," Jiang stated. After describing his efforts to do just that, Jiang added, "I believe more and more people will come back as our economy develops and research conditions improve."[83]

In the twenty-first century, more Chinese students have gone abroad than ever before. The United Nations Educational, Scientific and Cultural Organization (UNESCO) estimates that, from 2003 to 2013, the number of Chinese students at the tertiary level overseas jumped from around 313,000 to more than 712,000.[84] Yet China's leaders have also stepped up their efforts to lure students back to China. As early as 2004, China had established fifty-two educational bureaus in the thirty-eight countries where overseas students were most concentrated, and these helped form more than 2,000 Overseas Students Associations.[85] This overseas presence has helped Beijing inform Chinese students about developments and opportunities at home, while also monitoring the political activities of Chinese students abroad.

The CCP itself has become directly involved in the effort to recruit talent from abroad. In 2003, the CCP Politburo created the Central Coordinating Group on Talent (CCGT), which would be led by the Organization Department of the Central Committee with participation from a dozen different ministries.[86] In 2008, the CCGT unveiled the Thousand Talents Plan, designed to entice 2,000 highly qualified individuals to return to China over the next five to ten years.[87] There was particular interest in individuals who could "make breakthroughs in key technologies" or serve as leaders in newly emerging scientific fields. Subsequent initiatives have included the Thousand Youth Talents Plan of 2010, the Thousand Foreign Experts Plan of 2011, the Special Talent Zone of 2011, and the Ten Thousand Talents Plan of 2012.[88] Chinese high-tech talent has also been encouraged to return by the country's growing number of high-tech parks, which help returnees access capital and other resources, as well as by tax breaks and other incentives.[89] More recently, the Chinese government has made efforts to reform its scientific research funding system, creating new programs and promising reforms to empower scientists rather than bureaucrats.[90]

China's efforts to encourage return migration have met with some success. Although the number of students going abroad has increased markedly in the past decade, the number returning has risen even faster, as shown in figure 1.2. As a result, the ratio of returnees to students going out has risen. In 2007, the number of students returning was 31 percent of those

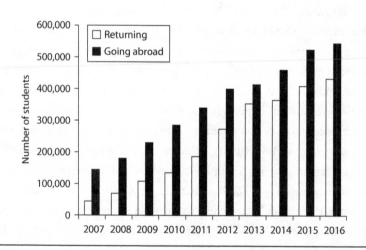

FIGURE 1.2 Number of Chinese students going abroad and returning, 2007 to 2016. *Source*: Ministry of Education of the People's Republic of China, "Liuxue Renyuan Qingkuang Tongji [Statistics on the Situation of Overseas Students]," *Ministry of Education of the People's Republic of China*, various years, accessed March 1, 2017, www.moe.edu.cn.

going abroad. By 2013, that figure had risen to 85 percent. Since then, it has hovered between 78 and 80 percent.

China's programs focused on attracting S&E talent in particular have also had some success—at least in quantitative terms. By 2012, 2,263 individuals had registered to return under the Thousand Talents program—exceeding the initial goal.[91] The Thousand Youth Talents Program—aimed at individuals under forty years of age—has also been successful. In 2011, the first group of 143 returnees was announced.[92] In 2015, 2016, and 2017, 661, 558, and 590 individuals returned, respectively.[93] Data from the United States also suggest a return flow of scientific talent to China. In 2001, 98 percent of Chinese students who had received Ph.D.s from U.S. universities in S&E disciplines five years earlier were still in the United States. By 2011, that figure had fallen to 85 percent.[94]

In many regards, however, China still has far to go. The country clearly struggles to attract non-Chinese talent: Only 381 individuals had been selected for the Thousand Foreign Experts Plan as of 2017.[95] In addition, the Thousand Talents Plan has struggled to bring back the topmost talent, particularly among academics, who are likely deterred by China's authoritarian political system, its bureaucratic research environment, and its lack of transparent decision-making.[96] Further, some scholars have fabricated foreign credentials in order to benefit from the program.[97] Such problems "could have long-term negative impacts on China's scientific and educational community by turning the best and the brightest away," notes Cong Cao, an eminent scholar of China's scientific elite. "They don't want to be in the company of shoddy academics, even if they make up only a handful."[98]

Because China's leaders appreciate at least some of the obstacles to reverse migration, they have also embraced the "diaspora model," as other developing countries have done.[99] In this view, members of the diaspora who do not return may still participate in national development through scholarly exchanges, business relationships, and other types of collaboration. In the 1990s, China explored this approach through the "spring light project," which brought overseas mainlanders back for short-term visits. In 2001, the Chinese government embraced the diaspora model more generally in a document co-authored by five ministries, which encouraged Chinese overseas to "serve the nation" even if they did not return to China.[100] More specifically, overseas Chinese were encouraged to do the following:

1. Make use of the advantages of their professional bodies;
2. Hold concurrent positions in China and overseas;
3. Engage in cooperative research in China and abroad;
4. Return to China to teach and conduct academic and technical exchanges;
5. Set up enterprises in China;
6. Conduct inspections and consultations; and
7. Engage in intermediary services, such as run conferences, import technology or foreign funds, or help Chinese firms find export markets.[101]

Nonetheless, China's enthusiasm for the diaspora model has its limits. Indeed, there is a certain uneasiness in China's attitude toward the prospect of greater opportunity for its students and professionals studying and working abroad. For example, the Chinese media sometimes describe U.S. work visas as a challenge for China. Chen Weihua, the deputy editor of *China Daily USA*, for example, has written that expanding employment-based visas and the H-1B visa program "would pose a huge challenge for China, which has been making great efforts to attract and retain talent."[102] There is evidence that the government agrees. In mid-2013, as the U.S. president, Barack Obama, and his allies in Congress were pressing for a comprehensive liberalization of U.S. immigration policy, the top official in China's central coordinating office for talent told the *People's Daily* that China must face challenges posed by other countries' immigration policies by redoubling its efforts. "We must maintain our momentum," the official said, "using even more flexible policies and mechanisms to participate in the international competition for talent."[103] This emphasis on competition for talent is an important theme in China. In fact, Chinese expert Wang Huiyao has written of a "global talent war"—one in which China is competing with the United States and other countries to attract the world's best and brightest—though such martial rhetoric is hardly unique to China.[104] In February 2015, Premier Li Keqiang made Wang a counselor to the Chinese State Council.

Going forward, China's leaders hope to see more impressive results from their policies. In 2013, President Xi Jinping stated his support for the liberal party line on study abroad adopted in 1993: "Support overseas study, encourage people to return, and give people the freedom to come and go." He also added four new characters to the twelve-character slogan, *fahui zuoyong* ("produce a marked effect"), suggesting hope and determination to see the liberal policy produce greater benefits for China in the future.[105] In May 2015, Xi reiterated this sixteen-character slogan and added that forming a "united front" with China's overseas students should be a priority.[106] Against this backdrop, the China Scholarship Council (CSC)—a nonprofit agency associated with the Chinese Ministry of Education—has stepped up its financial support for Chinese students going abroad in recent years. From 2012 to 2017, the number of scholarships planned by the CSC jumped from 16,000 to 32,500.[107] Scholarship funding normally requires students to return home after completing their studies, and in 2014, the Council reported that 98 percent of the students it had supported since 1996 had done so.[108]

China has come to adopt a liberal approach to the emigration of students and well-educated professionals in recent years, but it has also intensified its efforts to bring these individuals back to China. Such efforts have met with no little success, but China's current leadership clearly hopes to see greater results in the future.

Global R&D

China has opened to global R&D in a variety of ways. First, openness to foreign investment has long been a key element of China's reform and opening process, and openness to foreign R&D has emerged as a noteworthy part of this story. Even so, there were misgivings about having foreign R&D centers within China when multinationals first expressed interest in opening them. In 1998, when Microsoft first opened its research lab in Beijing, the minister of Information Industry declined an invitation to attend the opening ceremony, and suspicious local media warned that "the wolf is here."[109] More generally, there were concerns about multinationals

luring China's best and brightest away from domestic firms and institutes. At the turn of the century, articles in state-controlled newspapers
called well-educated Chinese who went to work for multinational firms
"traitors."[110]

In recent years, such concerns have receded. Many of China's leading
firms now invest impressive sums in R&D, and opportunities and salaries
at domestic companies have become more attractive.[111] Further, Chinese
startups have become much more prevalent and are also now competing
for talent.[112] More generally, China's leaders have clearly decided that the
country must deepen its integration with the outside world in order to
advance. This conviction is evident in key national documents. In 2006, the
National Medium- and Long-Term Program for Science and Technology
Development (2006–2020), or MLP, argued that China should increase
its collaboration with the rest of the world in science and technology.[113]
Universities and research institutes were exhorted to establish joint laboratories with foreign institutions, and multinational corporations were
invited to set up more R&D centers in China. In 2010, a circular outlining
a new effort—the Strategic Emerging Industries initiative—also stressed
the need for continuing international cooperation.[114] Like the MLP, it
also called for foreign companies to set up R&D centers in China and for
more foreign investment in key sectors. In 2011, the Twelfth Five-Year Plan
encouraged "foreign enterprises to establish R&D centers in China" so that
the country could "actively merge into the global innovation system."[115]
Various incentive schemes have also encouraged the formation of R&D
centers by both domestic and foreign firms.[116] The Chinese government has
even pressured foreign firms to set up R&D facilities in China in exchange
for market access.[117]

To be sure, disclosures regarding U.S. cyber-surveillance overseas have
fomented suspicion of U.S. technology firms in Beijing in recent years.
This has marred relations between these firms and the Chinese government and has had an impact on their business in China.[118] Even so, there
is still strong interest in luring foreign R&D (including that of U.S. firms)
to China within the Chinese government.[119] In January 2014, in fact, the
People's Daily rejoiced that China was "attracting key R&D activities from

multinational companies."[120] Even Google, which famously clashed with Beijing after it stopped censoring results on its China search engine in 2010, continued to conduct R&D in China thereafter.[121]

The nature of foreign R&D in China varies considerably. In the past, "asset-exploiting" R&D—R&D that adapts existing products to the Chinese market—has been seen as dominant.[122] In some cases, this involves making products less advanced in order to suit outdated local processes or to compete on price.[123] One prominent international wind power firm, for example, redesigned its gearbox in China to make it less expensive—but in doing so, the firm cut the durability of the product in half.[124] In other cases, however, reengineering products for the Chinese market produces ideas and technologies that are useful worldwide.[125] In other words, the line between asset-exploiting and asset-augmenting R&D has blurred.

There are also clear examples of asset-augmenting R&D. In fact, a 2008 survey identified fifty-one multinational firms carrying out "innovative" R&D (i.e., relevant to the firm's global R&D operations).[126] One of the best examples is Microsoft's high-powered research center in Beijing, which had 230 permanent R&D staff by 2014 and hundreds of visitors annually.[127] GE's China Technology Center in Shanghai, one of four multidisciplinary R&D centers that the company operates outside the United States, conducts advanced research in areas ranging from manufacturing technologies to materials science.[128] And Intel's China Research Center works closely with R&D teams in the United States, India, Korea, and Russia, though global product development is done elsewhere.[129]

Chinese firms, meanwhile, are expanding their R&D investment abroad. From 2007 to 2013, for example, R&D spending by Chinese firms in the United States grew rapidly from virtually nothing to $449 million.[130] This increasing activity has been encouraged by Beijing. In 2006, for example, China's MLP called for "encouraging and helping [Chinese firms] to establish R&D centers or industrialization bases overseas."[131] Chinese firms have been far more active, however, in investing in foreign technology companies. From 2005 to 2016, Chinese firms spent nearly $58 billion to gain stakes in or acquire foreign technology firms, with most of the activity occurring in recent years.[132] Beijing has often been supportive here as well.

In fact, it has been seen as *too* supportive: Beijing's support for Chinese firms' foreign acquisitions, particularly in semiconductors, has aroused concern in the United States.[133] As of 2017, the Chinese government was scrutinizing overseas investments more carefully, and Chinese activity was expected to subside—at least temporarily.

China has also begun to embrace cross-border R&D collaboration. Collaboration and co-authorship between U.S. and Chinese scientists have increased with remarkable rapidity. Yet there is also interest in more formal arrangements. The MLP, for example, encouraged "research institutes and universities to establish joint laboratories or R&D centers with overseas research institutes."[134] The circular announcing the Strategic Emerging Industries initiative, in turn, called for a greater number of "technology innovation alliances led by enterprises and joined by research institutes and universities," although this goal was not exclusively aimed at the outside world.[135] The Chinese government has also supported international R&D collaboration in other official documents.[136] This openness, in turn, has allowed China to become a more prominent source of partners in transnational R&D alliances, as shown in table 1.2. In some cases, such collaboration is designed to promote the localization of a foreign product in China. Qualcomm, for example, has concluded deals with a number of Chinese partners toward this end. In other cases, the aims are more ambitious. Life Technologies of San Diego and Da An Gene of Sun Yat-sen University in Guangzhou, for example, created a joint venture in 2012 to develop molecular technologies to assist in the early diagnosis of cancer, infectious diseases, and genetic diseases.[137]

China has also opened to foreign venture capital, though this is very much a work in progress. Foreign VC firms first came to China in 1992, with the entrance of IDG, but it took until 2001 for such firms to gain legal recognition in the country.[138] Despite various reforms since then, the domestic regulatory environment remains challenging. Within the ICT realm, for example, restrictions on the foreign ownership of Internet companies create headaches for foreign VCs. More generally, the government's management of initial public offerings and equity markets makes it difficult for VCs to exit their investments within China.[139] Government support for domestic VC funds also creates massive distortions in the market. Even so, prominent foreign VCs, including Sequoia Capital and Kleiner Perkins Caufield & Byers, are active in China, and they have played an important role in internationalizing Chinese startups, helping them list overseas, and introducing them to foreign bankers, accountants, and lawyers.[140] And more reform has been pledged: In 2016, the Chinese government promised foreign VCs equal treatment with domestic firms.[141]

The Chinese government has opened to many aspects of global R&D over the past few decades. After some initial misgivings, it has enthusiastically welcomed foreign R&D centers, as well as collaboration with foreign partners, albeit with more rapidity in some cases than in others. China's increasing openness in this regard is striking, particularly as foreign firms have encountered significant difficulties in the country in other respects in recent years.

INDIA

Like China, India has sought to harness global innovation in pursuit of its developmental goals. These goals appear to be quite ambitious. In 2013, the government unveiled its Science, Technology and Innovation policy with the aim of placing India "among the top five global scientific powers" by 2020.[142] In early 2017, Prime Minister Narendra Modi vowed that India would be "among the top three countries in science and technology" by 2030.[143] Yet India has approached global innovation rather differently than China has. India has not matched China's efforts to encourage reverse migration or attract foreign investment in R&D. Even so, India stands out with regard to how its leaders and companies have embraced the diaspora model.

Flows of Brainpower

India has a long history of sending well-educated individuals overseas. Large-scale migrations of laborers began after slavery was terminated in British colonies in the 1830s, with most migrants going to South or

Southeast Asia. Whereas most of these workers were unskilled, a second wave of migration of professional workers—clerks, bureaucrats, and traders—followed and continued in the first half of the twentieth century.[144] Following independence, however, the Indian government did not initially embrace the idea that its citizens had a right to go abroad. Prior to the mid-1960s, in fact, the government considered the issuance of passports to be a matter of its discretion in its conduct of foreign affairs. Only in 1966 did the Indian Supreme Court establish the "right to travel" as a fundamental right under the Indian constitution; this prompted the Indian parliament to enact the Passports Act of 1967.[145] Even so, the right to travel remains regulated: The Act allows the government to deny the issuance of a passport if the government believes doing so would not be in the "public interest."[146] Subsequently, the Emigration Act of 1983 stipulated that the government could prevent emigration by "any class or category of persons" if doing so were in the interest of the "general public."[147]

At least in theory, then, the Indian government has some legal authority to curb emigration. It can also discourage emigration through a variety of other means, ranging from taxation to public criticism, that would make going abroad more difficult.[148] In practice, it has not done so—but the reasons for that have changed over time.[149] Starting in the mid-1960s, Indian emigration to the United States accelerated, particularly among more educated citizens, and concern about brain drain grew. In the 1970s, the exodus of talent took on particular significance in the high-tech realm, as it became common for graduates of the elite Indian Institutes of Technology to migrate to the United States for further study and then to stay for work.[150] The government did not try to stem the outflow of human capital, in part because the old-guard political leadership did not wish to deny opportunities to other members of the country's social elite, particularly at a time when opportunities within the country were constrained by a weak economy. Newer political leaders, meanwhile, were happy to see members of the old elite leave the country. In other words, the brain drain acted as a safety valve for India at the time.[151]

Over time, India's rationale for allowing its best-educated citizens to go abroad became a more positive one. In the 1980s, Prime Minister Rajiv Gandhi described the Indian diaspora as a "brain bank"—an asset the country could call upon in its development—and this way of thinking began to catch on in the early 1990s.[152] Since then, prime ministers from both the Bharatiya Janata Party (BJP) and the Congress party have emphasized increasing interaction with educated Indians overseas, rather than trying to limit the outflow of brainpower.[153] It is against this backdrop that well-educated Indian workers have flowed in terrific numbers to other countries, particularly after the Internet boom began in the latter half of the 1990s. These flows had an impact on the emigration rate for Indians with a tertiary education, which jumped from 2.6 percent in 1990 to 4.2 percent in 2000.[154] By 2010, India had 2.2 million expatriates with tertiary educations in OECD countries. The cohort of Indians moving to the United States has been called the "IT generation," since so many of these individuals are employed in ICT industries. By 2012, in fact, one-quarter of India-born workers in the United States were employed in computer-related occupations. And this group continued to grow: By 2014, India had become the largest source of new immigrants and the second-largest source of total immigrants (after Mexico) in the United States.[155]

Indian students have been globe-trotters as well. In 1999, there were 55,670 Indians studying overseas, according to UNESCO. Ten years later, UNESCO's estimate had risen to 203,497, whereas the Indian government estimated that more than 250,000 were abroad at that point.[156] The most popular destination for Indian students has been the United States: In 2014/15, a total of 132,888 Indians were studying at U.S. higher education institutions.[157] In fact, India had more students in the United States than any other country did between 2001/02 and 2008/09. A sizable portion of Indian students in the United States are enrolled in graduate S&E programs. As of November 2014, for example, more than 72,000 Indian students were studying in American graduate S&E programs, with engineering and computer science programs enrolling the vast majority of these students.[158] India has also sent large numbers of students to other English-speaking countries. In 2012, nearly 30,000 Indian students were studying in the United Kingdom, and more than 40,000 were enrolled in Australia in 2013; however, these figures have fluctuated significantly in recent years.[159]

Like its Chinese counterpart, the Indian government has long hoped for return migration from its professionals and students abroad. As Rajiv Gandhi put it in 1985, "We must produce the atmosphere, the infrastructure, for (return) in industry, in our research establishments, in our academic establishments."[160] Toward this end, the government has created programs to encourage members of the diaspora to return, and these have produced some limited results.[161] Nonetheless, the scale and intensity of China's effort to induce return migration is not evident in India. There is no Indian analogue to the China Scholarship Council, an organization that supports overseas study but requires students to return home following graduation.[162] Nor is there an Indian counterpart to China's Thousand Talents Program, a high-profile and well-financed effort to bring home particularly capable members of the diaspora. This is understandable. The Indian government does not have the resources enjoyed by its Chinese counterpart, which presides over a much larger economy. Moreover, Indian firms' investments in R&D are but a fraction of what Chinese firms are spending, and Indian universities are not as ambitious as their Chinese counterparts. The domestic demand for high-end talent is thus not as acute as it is in China.[163]

Just how much return migration India has experienced remains unclear. Unlike the Chinese government, the Indian government does not publish data regarding the number of students returning from abroad. Data from the United States, however, suggest a strong propensity for many Indian individuals abroad to remain overseas. In the early 2000s, the rate at which the India-born population on temporary visas became permanent residents in the United States was very high, perhaps more than 90 percent.[164] Although this rate may have fallen in recent years, given anecdotal reports of increased reverse migration, the India-born population becoming long-term residents of the United States still appears to be growing by more than 100,000 per year.[165] And like China, India continues to struggle with bringing back its most talented students. In 2001, 86 percent of Indian students who had received a Ph.D. in an S&E field from a U.S. university five years earlier were still living in the United States.[166] By 2011, that figure had fallen to 82 percent.[167] Although the figure has fallen over time, it remains quite high and has fallen more slowly than in China's case. Recent research also suggests that the propensity of Indian students to return from the United States is negatively correlated with ability, post-migration education, and income.[168]

If India has been less preoccupied with return migration than China has, India's enthusiasm for the diaspora model is particularly striking. While India's relationship with its diaspora was tepid for decades after independence, it underwent an important shift around the turn of the century. In 1999, Prime Minister Atal Bihari Vajpayee's government created the High Level Committee on the Indian Diaspora. The Committee's extensive report ushered in what has been called "a paradigm shift" in India's relationship with its expatriate community.[169] Offering a range of recommendations, the report concluded that "the reserves of goodwill amongst [the Indian] diaspora are deeply entrenched and waiting to be tapped if the right policy framework and initiatives are taken by India."[170]

A variety of initiatives followed. In 2002, Vajpayee's government revised the Person of Indian Origin (PIO) scheme, reducing the cost to facilitate visits from some foreign individuals with Indian ancestry. In 2004, the government created the Ministry of Overseas Indian Affairs to increase engagement with the diaspora. In 2006, Prime Minister Manmohan Singh's government formally launched an expanded Overseas Citizen of India (OCI) scheme, which targeted individuals who had migrated from India and acquired citizenship in another country. In 2009, Singh's government created the Prime Minister's Global Advisory Council to facilitate communication between the Indian government and prominent members of the diaspora.[171]

In the past, observers of the Indian government's efforts have charged it with focusing on narrowly conceived incentive schemes while neglecting broader economic reforms that could attract greater interest from the diaspora—and rightly so.[172] Under Narendra Modi's government, this has changed. Modi has consolidated previous efforts, particularly by moving the Ministry of Overseas Indian Affairs into the Ministry of External Affairs and by merging the PIO and OCI schemes, while promoting his efforts to liberalize the Indian economy to the diaspora. In September 2015, for example, Modi addressed a crowd full of Indian expatriates in Silicon Valley.

Modi said well-educated Indians abroad were not a brain drain but a "brain deposit" and that these "deposited brains" could serve their homeland through a variety of means "whenever the opportunity comes."[173] Touting his reformist record and credentials, Modi stated that the opportunity to serve "Mother India" had arrived.

India's positive attitude toward its diaspora, including its overseas IT workers, reflects multiple considerations. First, Indian officials have long sought to attract financial inflows from Indian workers abroad, and such inflows have increased considerably in recent decades, particularly in the form of remittances.[174] Although not all of these funds come from professional workers, the large number of such workers abroad has clearly helped boost the national total. For 2014, in fact, the World Bank has estimated that India received more than $70 billion in remittances—more than any other country.[175] The total was equivalent to 3.7 percent of India's GDP and 23.5 percent of its foreign exchange reserves.

Even more relevant here, India's most prominent technology companies have come to rely on a business model predicated on the existence of a substantial diaspora of well-educated workers. This is particularly evident in the United States, where Indian companies rely on the H-1B visa program to export their services to American customers. Since the 1990s, such companies have used this program (and, to a lesser degree, the L-1 program for intra-company transfers) to send Indian computer programmers and software engineers to locations in the United States, and it has become a vital part of their corporate strategies.[176] In 2014, in fact, seven Indian firms—led by Tata Consultancy Services, Infosys, and Wipro—received 16,573 of the 85,000 private sector H-1B visas allocated.[177] These overseas employees then allow the Indian firms to provide high-tech labor to customers in North America, which is by far their largest market.

Thanks to these dynamics, the Indian government's perspective on U.S. immigration policies differs significantly from that of its Chinese counterpart. Whereas China has traditionally worried about a more open U.S. system, India has been more worried about the opposite. Since the late 1990s, in fact, India has pressed the United States to increase the number of H-1B visas that it makes available and to make it easier for Indians to work in the United States.[178] In addition, when the U.S. government has created impediments to this flow, India has protested, even threatening action against the United States at the World Trade Organization.[179] In 2015, Prime Minister Modi personally voiced concerns about limits on the H-1B visa program when U.S. President Barack Obama visited India.[180] The Indian government has been much less concerned with employment-based visas, since Indian firms rarely seek green cards for their employees in the United States.[181] To be sure, some Indian commentators worry that a surge in U.S. employment visas, particularly for foreign graduates in S&E fields, could induce more Indian students to go abroad, complicating India's efforts to become a scientific power in its own right.[182] At least until 2014, however, the topmost officials in the Indian government generally had not been concerned with the prospect of a greater number of employment-based visas in the United States when such changes were mooted.[183]

India has traditionally adopted a liberal policy toward cross-border flows of brainpower, and the number of Indian students and professionals abroad has reached new heights in recent decades. Unlike China, however, India has not invested great sums to bring back its overseas talent, and for a number of reasons, the country is more comfortable relying on the diaspora model.

Global R&D

Like China, India has embraced global R&D, but the process has been a gradual one. In the early 1970s, the chair of Texas Instruments, Patrick Haggerty, visited India and proposed setting up a manufacturing and research facility for integrated circuits. Although Indira Gandhi was open to the idea, she doubted it would pass muster in the political system more generally.[184] In 1985, Texas Instruments would try again, and this time it would succeed, setting up a design center in Bangalore. Since then, Texas Instruments has been joined by many other prominent multinationals that have established R&D centers in India, and Bangalore in particular has become a prominent hub for such operations. GE's Jack F. Welch Technology Center in Bangalore, to give one example, employs more than

5,300 technologists, making it the company's largest lab outside the United States.[185] Indeed, although India has not been as open as China to foreign direct investment (FDI) in general, it has been relatively open in the realm of informatics, including software development and R&D, thanks in part to connections to the outside world provided by the Indian diaspora.[186] One official study identified 964 foreign investments in R&D facilities in India between 2003 and 2009 with a collective value of more than $29 billion, about 8 percent of India's total FDI during that time, with most of the activity in ICT.[187] By 2016, according to the consulting firm Zinnov, India was home to 1,208 foreign R&D centers established by 943 multinationals.[188]

The nature of foreign R&D in India is diverse, including both asset-exploiting and asset-augmenting activity. Past research has argued that, compared with China, foreign firms had a greater tendency to conduct asset-augmenting R&D in India, particularly as multinationals offshored elements of their R&D activities to Bangalore and other Indian locations.[189] The IBM Research labs in Beijing and Bangalore offer a good example of this tendency: The Beijing lab is focused more on developing products suitable for the Chinese market, whereas the Bangalore lab is involved more in IBM's global R&D effort.[190] This general tendency may no longer be the case, or as much the case as it once was, as foreign firms have begun investing in more global-facing R&D in China in recent years. In any case, the line between asset-exploiting and asset-augmenting R&D in India is blurry, as it is in China, as products developed for the Indian market may also be marketed globally. In particular, multinationals have developed less expensive versions of developed-world products in India, relying on "frugal engineering," subsequently marketing these products worldwide.[191]

India has also cultivated international collaborations. India's 2013 Science, Technology and Innovation policy stated that the country would pursue "strategic partnerships and alliances through both bilateral and multilateral cooperation."[192] In addition, the policy noted the high cost of "global R&D infrastructure" and thus encouraged participation in "international consortia." Such collaboration would allow "Indian industry to gain global experience and competitiveness in some high-tech areas." The government's stated interest in technology alliances is backed up by the fact that a wide range of prominent Indian firms and organizations have collaborated with foreign partners through such arrangements.[193] Indian participants include Dr. Reddy's Laboratories, Glenmark, the Indian Space Research Organisation, Infosys, Lupin, Mahindra & Mahindra, Reliance, and Tata Sons. If one includes software development and computer programming alliances, the list of Indian participants also includes HCL, Tata Consultancy Services, and Wipro, among other companies.

Under Prime Minister Narendra Modi, India has become particularly enthusiastic about collaborating with multinationals in R&D. To be sure, the government's top priority is attracting foreign investment in manufacturing, which has greater potential to generate employment for most of the country's growing population, as evidenced by the "Make in India" campaign.[194] Even so, India's leaders do wish to expand the country's role in global R&D. "There is a growing trend of international collaboration in research and development," Modi noted in January 2015. "We should take full advantage of this."[195] Modi's speech in Silicon Valley in September of that year praised the local Indian community's role in high-tech innovation and exhorted it to play a greater role in India's development.[196] If manufacturing is India's top priority, then India's leaders are eager to raise the country's profile in global R&D as well.

India has also come a long way in opening to foreign VC over the past several decades. In 1993, the U.S. investor Vinod Khosla began commuting between Silicon Valley and India, hoping to develop a VC business in India, only to give up after three years.[197] By 1999, many foreign VCs had established a presence in India, and roughly 80 percent of all VC investments in India were derived from foreign firms. Still, a variety of regulatory and legal barriers to startup firms and VC investment remained. In more recent years, the situation has greatly improved. The Indian startup scene has become far livelier over the past decade, inspired by such success stories as Flipkart and Ola. Foreign VC firms, meanwhile, hoping to find the "next big thing," have taken greater interest in India since Alibaba's $25 billion initial public offering in 2014.[198] The Modi government has worked to encourage both of these trends. Modi's "Startup India" initiative has created incentives and increased funding for new firms, though it remained a work in progress as of mid-2017.[199] The government has also become more welcoming to foreign VCs, allowing them to invest in startups regardless of sector without prior permission from the central bank.[200]

India has opened itself considerably to global R&D over the past few decades. Although the Indian government has not courted it as eagerly as China has, foreign R&D in India has grown considerably. India has also sought greater international collaboration through R&D alliances and VC investments.

CONCLUSION

The globalization of innovation seen in recent decades is a development without parallel in history. Skilled workers and students are increasingly mobile, and multinational firms and research universities are engaging in more cross-border R&D investment and collaboration than ever before. The world's most prominent rising powers, meanwhile, have embraced both of these developments in their efforts to emerge as innovation leaders in their own right. Such openness does not reflect disregard for the dangers, however. On the contrary, both China and India have come to their current level of enthusiasm for global innovation gradually and with occasional second thoughts. To varying degrees, Chinese and Indian leaders have worried about whether the talent they send overseas will return, and there have been concerns about foreign R&D centers attracting the top domestic talent as well. In both cases, then, the potential dangers are understood, but the mainstream view is that these are outweighed by the potential benefits of openness.

China and India have approached global innovation in different ways. The Chinese government is more purposeful and activist, encouraging the country's overseas talent to return and working to attract foreign investment in R&D. India's approach, by contrast, seems more ad hoc. And although Indian leaders are certainly interested in reverse migration, India is more comfortable with—and more reliant on—the diaspora model. In both cases, however, there is hope and conviction that the country will profit from global innovation.

These developments raise important questions for many countries, but particularly for the world's dominant state and technology leader. In the chapters that follow, I consider the forces that shape how the United States participates in global innovation.

2

INNOVATION LEADERSHIP AND CONTESTED OPENNESS

I F FOREIGN ECONOMIC policy is important, that of the world's most powerful state is particularly so. It comes as little surprise, then, that the forces behind the leading state's foreign economic policy have attracted so much interest over the past several decades.[1] Theorists focused on the international position of the hegemon have depicted it as a force for openness and order in the world economy, either as a provider of a public good or as a self-interested welfare-maximizer.[2] A more state-centered strand of research focuses on the interests and authority of national policymakers or the persistent influence of national institutions.[3] Others highlight the role of economic ideas, including those that took root before the state in question became dominant.[4] There is also an extensive literature on the role of societal interests in shaping U.S. foreign economic policy.[5] And some works weave two or more of these strands together.[6]

Focusing on policies toward global innovation in particular, the theory I develop in this chapter combines several distinct strands of scholarship. In the first section, I highlight the position of the dominant state as the innovation leader in the international system, building on the work of other scholars. The discussion here focuses on the dominant state's preeminence in "leading sectors" in particular, and it explains how such preeminence supports the hegemon's primacy. In the second section, I document the evolution of industrial innovation in the dominant state over the past two centuries and, in particular, the emergence of the high-tech community (HTC) in the United States in recent decades. In the third section, I then consider what the development of the HTC implies for the dominant state's policies toward global innovation. Specifically, I develop a theory focused on shared interests within the HTC and the political competition between the HTC and different types of opponents. I also outline several alternative explanations for the dominant state's policies toward global innovation. In the final section, I explain how the theory will be tested in the chapters that follow.

INNOVATION AND HEGEMONY

As Joseph Schumpeter noted many decades ago, innovations are not randomly distributed in time and space, but instead have a tendency to cluster together.[7] Among international relations scholars, Robert Gilpin was among the first to apply this insight, noting the key role that "technological revolutions" played in the rise to preeminence of the United Kingdom and the United States.[8] Gilpin also argued that over time, technological know-how and "inventiveness" would diffuse to other countries, leading to the decline of the dominant power and disrupting the established order.[9] More recently, "long-cycle" theorists have explored the relationship among innovation leadership, economic dynamism, and global primacy with particular interest.[10] These theorists argue that new countries become dominant because they develop a series of innovations in "leading sectors." These are relatively new, high-growth sectors that generate positive spillovers and drive wider changes in the national and world economies.[11] In recent years, other scholars of international relations have emphasized how innovation in leading sectors supports hegemonic power.[12] Although the term *leading sector* is not typically used in other disciplines, the concept frequently appears under a variety of other names.[13]

Over the past two centuries, the locus of innovation in leading sectors has shifted from the United Kingdom to the United States. In the early phase of the industrial revolution, the United Kingdom led in steam engine and railroad technology. In the latter phase of the industrial revolution, the United Kingdom ceded leadership in steel, chemicals, and electricity to the United States and Germany. In the mid-twentieth century, the United States predominated in automobiles, aviation, and electronics. In the latter years of the twentieth century and early in the twenty-first century, the United States has led the way in information and communications technologies (ICT).[14]

The clustering of leading-sector innovation in the United States today is worth highlighting. Although the ICT revolution has hardly been limited to North America, the concentration of leading ICT firms and related academic expertise in the United States has not been replicated in any other country.[15] In recent years, the United States has accounted for less than 5 percent of the world's population and a quarter of the population of member countries of the Organisation for Economic Co-operation and Development (OECD).[16] In 2016, however, the United States was home to sixty-five of the world's top one hundred software companies, as measured by R&D spending.[17] No other developed country had more than five. That same year, U.S. firms accounted for fifty-six of the world's top one hundred ICT hardware companies. By comparison, Taiwan had thirteen, and Japan had ten. Overall, ICT firms accounted for 41 percent of all business R&D in the United States in 2013, a share exceeded in only four other economies among the thirty-six tracked by the OECD.[18] The top relevant universities are also clustered in the United States. For 2014/15, the *Times Higher Education* rankings listed fourteen U.S. universities among the top twenty for engineering and technology.[19] In short, the United States plays a unique leadership role in ICT.

If leading-sector innovation is concentrated in the dominant state, how does it generate and sustain primacy? First, and most basically, innovation represents a key source of economic growth. Particularly since the 1990s, economists have focused on innovation as a fundamental driver of national prosperity.[20] Whereas increasing inputs of capital and labor face diminishing returns, innovation increases the efficiency with which such inputs are used and thus represents a sustainable source of long-term growth.

Leading-sector innovation in particular has powerful implications for growth. Great Britain, for example, was already powerful in the eighteenth century, but the technologies of the industrial revolution propelled it to global preeminence in the nineteenth century. This was not because the United Kingdom became the world's largest economy—it did not—but because steam and railroad technology made it the world's most advanced productive power.[21] By 1860, with only 2 percent of the world's people, and only 10 percent of the European population, the United Kingdom possessed 40 to 45 percent of global capacity in modern industry.[22] Today, the ICT revolution is still unfolding, but it has already boosted the United States' productivity advantage over Europe.[23] It is also clear that ICT is becoming increasingly important to industrial innovation in the United States, as non-ICT firms are focusing a growing share of their own R&D spending on incorporating software and high-tech services into their products.[24]

The second key reason innovation sustains primacy is that new technologies often have military and intelligence applications. The automobile, the airplane, and information technology are several obvious examples from recent leading sectors. While new technologies, particularly those that are militarily useful, generally diffuse throughout the international system, the pace of diffusion varies widely. Some new technologies are terrifically expensive, some are difficult for military organizations to adopt, and some present both of these challenges.[25] In such cases, the pace of diffusion can be decidedly slow. There is no guarantee, of course, that a state that invents a new technology will be the first to exploit its potential in the military sphere. Nonetheless, states that invent and come to dominate the production of a new technology enjoy considerable advantages over their rivals. These include economies of scale, lower production costs, and greater expertise in further development and production.[26] It is telling that the United States, the ICT industry leader, has led the world in applying such technologies to military purposes.

In short, the economies of dominant states are unusual. Such states are innovation leaders: They predominate in the leading sectors that unleash Schumpeter's "creative destruction" on the national and world economies, and such leadership underpins their national power.

THE EMERGENCE OF THE HIGH-TECH COMMUNITY

If dominant states have traditionally specialized in leading-sector innovation, the actors responsible for producing these innovations have changed dramatically over the past two centuries. In eighteenth-century Britain, innovation was typically led by individual inventors who were frequently building on the work of other individuals. In the latter years of the nineteenth century, organized industrial research became increasingly common, particularly in Germany and the United States. Today, leading-sector innovation is dominated by ICT firms that invest heavily in R&D and research universities based in the United States.

A BRIEF HISTORY OF INDUSTRIAL INVENTION

British leadership in the first industrial revolution sprang from the genius of individual inventors. The quintessential technology of the early industrial revolution, the steam engine, was invented by a string of inventors in the eighteenth and nineteenth centuries, including the remarkable James Watt, whose role in power technology has been likened to that of Newton in physics. Other inventors, such as Henry Bessemer, loomed large in the invention of steel. In fact, it was two amateur British inventors who devised a way to make steel from the phosphorous-rich iron found widely in Europe. Similarly, one of the key breakthroughs that led to the birth of the chemical industry came from a British chemist, William Perkin, who discovered an artificial dye by accident in 1856, inspiring many others to follow in his footsteps.[27]

More organized and systematic research took shape in the second half of the nineteenth century, particularly in the German chemical industry.[28] Several German chemical firms, including Bayer and BASF, were formed in the 1860s. The firms often had academically trained chemists among their founders, and those that did not quickly brought one on board. Between 1877 and 1886, the seven largest chemical dye firms in Germany established dedicated laboratories. By 1890, the three largest German firms employed

350 academically trained chemists. This would soon catch on elsewhere. In 1902, DuPont established its Eastern Laboratory in New Jersey with a mission to improve the company's high-explosives products through scientific research.[29] Universities, in turn, became important sources of human capital for chemical firms.[30]

Electricity emerged as another arena of organized research in the nineteenth century. Thomas Edison's "invention factory" in Menlo Park, New Jersey, was completed in 1876, and Edison soon had a team of inventors working there under his charge.[31] Edison went on to form the Edison Electric Light Company in 1878 and, fourteen years later, helped found General Electric (GE). In 1900, GE established the General Electric Research Laboratory in a barn that served as a garage for one of the lab's founders. This was the first industrial research facility in the United States.[32]

Industrial R&D became more established in the United States in the decades leading up to World War II.[33] Chemicals were the primary site of activity. Chemicals and chemical-related industries (such as petroleum) accounted for nearly 40 percent of the industrial laboratories founded in the United States (and a comparable share of research scientists and engineers employed in the country's manufacturing industries) between 1899 and 1946. Over time, the electrical machinery and instruments industries became important employers of research scientists and engineers as well. In both cases, universities played a key role in supporting industry, with state universities paying close attention to the needs of the local economy, even though the United States was not yet a leader in academic research. This growing capacity for industrial R&D became evident during World War II. During World War I, the U.S. military had dominated R&D and production for the war effort, except in munitions (for which it relied on DuPont). During World War II, however, the U.S. government outsourced a much wider variety of military R&D and production, and a new organization—the Office of Scientific Research and Development—was established to oversee this more diverse effort, including government contracts with private firms and universities.

Following World War II, industrial research became considerably more prominent and widespread in the U.S. economy. Whereas in 1946, there were fewer than 50,000 scientists and engineers employed in industry, by 1962, this figure had jumped to roughly 300,000.[34] New leading sectors came into their own during this period, including automobiles, electronics, and aviation. Indeed, transportation equipment—including both autos and aviation—emerged as one of the five most research-intensive industries after World War II.[35] Accordingly, whereas DuPont had stood out for its R&D capability in the early twentieth century, new firms now came to the fore. In 1955, for example, General Motors (GM) introduced the Chevrolet small-block V8 engine, a revolutionary innovation that combined considerable power with light weight, compact size, durability, and simplicity.[36] That same year, GM became the first company in history to report a profit of more than $1 billion, topped the first list of the Fortune 500, and had its chief executive named "Man of the Year" by *Time* magazine.[37] In 1957, Boeing began production of the 707—the first commercially successful jet aircraft for civilian transport—a move that would cement the company's leadership of the aviation industry.[38]

In important ways, however, the HTC as we now know it had yet to emerge by this time. For one, the main source of R&D spending nationwide was the U.S. government. Thanks to a postwar boom, federal government spending on R&D exceeded business spending throughout the 1950s, the 1960s, and much of the 1970s. At its (relative) peak in 1964, federal spending represented 67 percent of total U.S. R&D.[39] Moreover, federally funded R&D during this period often had significant spin-off potential, particularly as military technologies (such as the jet engine and microelectronics) were adapted for commercial use.[40] Businesses and universities were the leading performers of R&D during this period, but not the main funders. Second, high-tech firms also relied heavily on the government as a vital consumer. Until the late 1960s, the U.S. government was often the preeminent, and perhaps the only, customer for the most advanced technologies in aviation, semiconductors, and computer hardware.[41] In short, it was a partnership: Industry and academia performed high-tech innovation, and the government was a vital funder and consumer.

In the last quarter of the twentieth century, high-tech interests in the United States became far more independent. The impetus for this was the arrival of a new leading sector: ICT. ICT companies had already existed for many years by that point: Some "old-economy" firms, such as IBM and AT&T, had their roots in the nineteenth and early twentieth centuries. From the late 1940s until the early 1970s, however, the invention of the transistor, the integrated circuit, and the microprocessor set the stage for radical new technologies and a revolution in industrial leadership. Although IBM helped pioneer the personal computing revolution, announcing the personal computer (PC) in 1981, the iconic firm would ultimately be eclipsed by such "new-economy" leaders as Apple, Cisco, Intel, and Microsoft. The newer firms would also preside over the merging of information and communications technologies starting in the late 1980s, particularly with the creation of the Internet, giving rise to "ICT" as the label for this new leading sector.[42]

The independence of ICT reflected two key changes. The first was the emergence of industry as the dominant funder of R&D nationally. In the 1980s, industry consistently funded half of the country's R&D, while the government share fell to 40 percent by the end of the decade.[43] With the boom in the ICT industry in the 1990s, industry's share soared, reaching nearly 70 percent of national R&D spending in 2000, while the government share fell to a quarter of the total. R&D-intensive ICT companies, including Intel and Microsoft, helped propel this shift.[44] By 2016, seven U.S. ICT firms were among the top twenty R&D spenders worldwide: Amazon, Alphabet, Intel, Microsoft, Apple, Cisco, and Oracle.[45] Universities played an important supporting role in this regard, providing human capital, anchoring innovation clusters in Silicon Valley and elsewhere, and shouldering more of the burden in funding basic research. Partly because of this last shift, the federal government's share of basic research funding fell from 71 percent in 1964 to 47 percent in 2013.[46]

The second key change was the decline of the U.S. government as a market for high-tech products. Whereas the government, particularly the Department of Defense, was initially the main customer for advanced ICT products, the commercial marketplace quickly became dominant. In the mid-1960s, for example, defense-related organizations were the main customers for integrated circuits in the United States, accounting for 70 percent of sales, but by 1980, their share had fallen to 10 percent.[47] The ICT industry's shift toward the commercial marketplace, in turn, influenced military procurement. In the mid-1990s, the Department of Defense began making greater use of "commercial off-the-shelf" purchases to avoid the growing cost of purchasing government-specific systems. As a result, ICT technologies developed for commercial users became more prevalent in the U.S. military, and innovation designed for the commercial marketplace became a more important component of military innovation.[48] This shift has resulted in a new dynamic between the public and private sectors. Whereas the government was once the unavoidable customer, the secretary of defense now travels to Silicon Valley in hopes of convincing high-tech firms to work with the U.S. military.[49]

The point here is not that the U.S. federal government has become unimportant in high-tech innovation. The government has adapted to the changing innovation landscape, creating new models for government–industry collaboration.[50] It remains the single largest investor in basic research, a crucial supporter of high-risk seed projects, and the largest consumer of information technology in the world.[51] Particularly for established ICT companies, however, the terms of collaboration have changed. Google has curtailed its work with the Department of Defense on robotics to focus on commercial uses, for example, while Apple famously refused to unlock an iPhone for the FBI.[52] More generally, some Silicon Valley firms shy away from working with the U.S. military for fear of losing out in the Chinese market.[53] In short, high-tech innovation in ICT remains a partnership among government, industry, and academia, but the terms of the partnership have changed, and the importance and autonomy of nongovernment actors have grown greatly.

THE HIGH-TECH COMMUNITY IN WASHINGTON

Whereas the high-tech community in ICT emerged as an important economic and technological phenomenon in the United States within the last quarter of the twentieth century, its representation in the nation's capital is another matter. In general, we can think of the HTC as consisting of two wings—corporate and academic—each of which has developed its own presence in Washington.

On the corporate side, leading ICT firms have frequently established their own representative offices in the capital.[54] IBM set up a public policy office in 1975, following years during which the company's leaders handled relations with Washington personally.[55] Intel followed suit in 1986. Apple established a small office in Virginia in the early 1980s, moved to a true Washington office in 1989, closed the office briefly in the 1990s, and then reopened it in 1999.[56] Microsoft was rather belated, setting up its Washington office in 1995, even though it first came under antitrust investigation in 1991. Cisco's Washington office was established in 1998. Younger firms like Amazon, Google, and Facebook set up their Washington outposts in 2000, 2005, and 2009, respectively. In addition, many ICT firms retain professional lobbyists or law firms to represent them in Washington, either as an alternative or in addition to having their own offices, whereas others manage relations with the federal government from their headquarters.

As their presence in Washington has increased, ICT firms have also greatly increased their lobbying budgets. Established firms such as IBM, Intel, Microsoft, and Oracle were already spending considerable sums by the late 1990s (table 2.1). In recent years, these firms have been joined by Internet companies, particularly Amazon, Facebook, and Google. In fact, Google became the top lobbying spender of all U.S. companies in 2012, spending $18 million, after which it fell back to third place in 2016.

ICT industries have also become far more active in contributing to political campaigns. Indeed, electronics manufacturers, computer software firms, and Internet companies have all become more active, particularly in the twenty-first century, as shown in table 2.2. A relatively small share of these contributions has come from political action committees (PACs) to federal candidates. In contrast, most of these contributions have come from individuals employed in the industry to federal candidates or have taken the form of "soft money" or "outside spending."[57] In 2016, for example, the $86.3 million contributed by electronics manufacturers included $41 million in individual contributions to candidates and $38 million in outside spending. Contributions from the electronics manufacturing industry typically favored

TABLE 2.1 LOBBYING EXPENSES OF LEADING U.S. ICT FIRMS: 1998 AND 2016

FIRM (R&D SPENDING 2016, $ BILLION)	LOBBYING EXPENSES ($ MILLION)	
	1998	2016
1. Amazon (12.5)	0	11.4
2. Google/Alphabet (12.3)	0	15.4
3. Intel (12.1)	1.1	4.2
4. Microsoft (12.0)	3.9	8.7
5. Apple (8.1)	0.2	4.7
6. Cisco (6.2)	0.6	2.0
7. Oracle (5.8)	1.9	8.6
8. Qualcomm (5.5)	0.4	5.6
9. IBM (5.2)	5.6	4.0
10. Facebook (4.8)	n/a	8.7

Note: The R&D spending data are based on the most recent full-year figures reported prior to July 1, 2016. See Barry Jaruzelski, Volker Staack, and Aritomo Shinozaki, "2016 Global Innovation 1000 Study," *PwC*, 2017, www.strategyand.pwc.com/innovation1000; Qualcomm, *Qualcomm Annual Report 2016* (San Diego, CA: Qualcomm, 2016), 15, http://investor.qualcomm.com/annuals-proxies.cfm.
Source: Center for Responsive Politics, "Lobbying Database," *OpenSecrets.org*, accessed March 3, 2017, www.opensecrets.org/lobby/.

Republicans in the 1990s but began to favor Democrats starting in 2004. The software and Internet industries have normally (if not always) favored Democrats, but the margin was generally not great until 2008. The apparent preference for Democrats in ICT reflects the importance of contributions from individual donors as opposed to PACs; corporate PACs in ICT have often favored Republicans.[58]

Last, ICT firms have also established a variety of business associations and other groupings to represent them.[59] Some of these emerged from earlier industrial groupings. The National Association of Office Appliance Manufacturers, first set up in 1916, became the Information Technology

TABLE 2.2 POLITICAL CONTRIBUTIONS FROM THE ICT INDUSTRY
($ MILLION)

YEAR	ELECTRONICS MANUFACTURERS	COMPUTER SOFTWARE	INTERNET
1990	2.9	0.3	0
1992	7.5	1.5	0
1994	6.3	0.8	0
1996	12.5	2.0	0.2
1998	11.7	3.8	0.1
2000	34.5	14.4	6.6
2002	27.8	14.3	2.1
2004	31.4	12.9	3.6
2006	22.4	9.2	2.5
2008	43.6	17.5	8.2
2010	26.0	10.6	4.3
2012	57.6	25.9	17.4
2014	28.1	12.9	11.9
2016	86.3	46.3	30.7

Note: These data are based on contributions of $200 or more from PACs and individuals to federal candidates and from PACs, soft money (including funds directly from corporate and union treasuries), and individual donors to political parties and outside spending groups, as reported to the Federal Election Commission. Note that a few firms have been assigned to more than one of the three categories in the table, so the totals should not be combined. *Source*: Center for Responsive Politics, "Interest Groups," *OpenSecrets.org*, accessed March 3, 2017, www.opensecrets.org/industries/.

Industry Council in 1994. The West Coast Electronics Manufacturers Association, established in 1943 to secure defense contracts, became the American Electronics Association (AEA) in 1978. The Association of Data Processing Service Organizations, established in 1960, became the Information Technology Association of America (ITAA) in 1991. In 2009, the ITAA and AEA merged, along with other groups, to become TechAmerica, which was acquired by another industry group in 2014.

Other ICT groups have more recent origins. The Semiconductor Industry Association, initially headquartered in Silicon Valley, was established in 1977. The Business Software Association was created by Microsoft in 1988 to fight software piracy. The group later became the Business Software Alliance and then "BSA | The Software Alliance." The Computer Systems Policy Project, founded in 1989, later became the Technology CEO Council. It consists exclusively of chief executives of prominent ICT firms, as the name suggests, and is typically chaired by one of them. TechNet, a grouping of high-tech and venture capital firms, was established in 1997.

On the academic side of the community, there is another array of associations. The most prominent is the American Council on Education (ACE), an umbrella organization that includes not only universities but also other higher education associations as members. More generally, ACE is one of the "Big Six" higher education groups in Washington. The other five include the Association of American Universities (AAU), the Association of Public and Land-Grant Universities (APLU), the American Association of State Colleges and Universities, the American Association of Community Colleges, and the National Association of Independent Colleges and Universities.[60] ACE has also coordinated meetings of the Washington Higher Education Secretariat, which includes roughly fifty higher education–related groups as members.[61] There are also advocacy groups with a particular interest in international students. The National Association for Foreign Student Affairs, which represents international education professionals, has been highly active in supporting flows of foreign students. And the Institute of International Education engages in a range of activities, including advocacy, research, and training, to promote international education.

Research universities are represented in a number of ways. Of the Big Six associations listed above, the AAU and the APLU are known for representing research universities in particular. As of 2017, the AAU included sixty-two leading public and private research universities, while the APLU included 235 public research universities and other members. Whereas the AAU's lobbying budget is relatively modest, the APLU has typically invested between $400,000 and $600,000 annually in federal lobbying expenses. In addition, the Council on Governmental Relations represents and advises research universities, medical centers, and research institutes conducting at least $15 million annually in federally sponsored research. Major research universities frequently have Washington offices and conduct substantial lobbying efforts as well. Harvard and Yale Universities, for example, spent $550,000 and $500,000 on federal lobbying in 2016, respectively, while Princeton spent $280,000, and Stanford spent $250,000.[62] Last, research universities have also collaborated in lobbying with large ICT firms, both formally and informally, as described in the chapters that follow.

THE HTC AND CONTESTED OPENNESS

What does the emergence of the HTC in the world's dominant state imply for how it approaches global innovation? To answer this question, we must turn to theory. First, we must theorize to what extent the HTC possesses a set of shared interests regarding global innovation that could form a basis for collective action. Second, to the extent that such common interests exist, we must ask when the HTC will succeed in pursuing these interests, assuming the state in question possesses a pluralist, liberal democratic political system. Third, we must consider alternative theoretical approaches to explaining policies toward global innovation.

SHARED INTERESTS IN THE HTC

What are the shared interests in the high-tech community with regard to policies toward global innovation? As other scholars have noted, there are a wide range of interests within the business community, even within a given industry, let alone a community that includes both ICT firms and universities.[63] The ICT industry itself is hardly monolithic: Its leading firms sometimes pursue conflicting agendas in Washington, targeting each other rather than pursuing some collective goal.[64] It is important, then, to explain why ICT firms and research universities would possess a set of shared interests with regard to policies shaping global innovation—and also to document what these interests are. To begin, let me specify that I am particularly interested in large ICT firms that invest heavily in R&D. Such firms are most likely to have a strong interest in the international mobility of capital and labor, to operate on a global scale, and to lobby the federal government. Smaller ICT firms may have similar interests, but their preferences may not be as strong, or they may be less well positioned to pursue them. With this consideration in mind, let us consider three key policy arenas in particular: foreign skilled labor, foreign students, and global R&D.

Let us first consider policies toward professional skilled labor. Previous studies have stressed the preference of "high-skilled capital" for expansive policies toward skilled immigration in liberal democratic regimes.[65] We can expect large ICT firms investing in R&D to be particularly strong advocates. Skilled immigration increases the overall labor pool and thus the supply and diversity of talent available to conduct R&D. This improves access to the most talented and creative workers, increases the chances that specific needs will be met, and promises to increase competition for jobs and restrain wage growth. This is important, since ICT firms' R&D is particularly labor intensive. In 2013, labor costs accounted for 76 percent and 67 percent of the domestic R&D costs incurred by U.S. ICT firms in services and manufacturing, respectively.[66] By comparison, the figures for the pharmaceuticals and aerospace sectors were each less than 40 percent. ICT firms' demand for skilled immigration, in turn, should be reinforced by demand from research universities. Such universities are also human-capital intensive, and they also desire domestic employment opportunities for their foreign graduates—opportunities that help them attract foreign students in the first place. The intensity of ICT firms' preference for openness is likely to vary over time, of course, depending on the fortunes of the industry. High-growth periods likely intensify demand for labor and the interest in openness, whereas downturns presumably have the opposite effect. Over time, however, both the corporate and academic wings of the HTC are likely to be strong supporters of openness toward skilled immigration.

Let us turn to policies toward foreign students. Universities have the most obvious interest in accessing a larger pool of such students. It is frequently noted that such access increases university revenues, either because foreign students enlarge the student body or because foreign students are charged higher fees.[67] Research universities have a particularly clear interest. In these cases, the admission of foreign students not only expands the pool of applicants at the undergraduate level but also enables the university to enroll greater numbers of talented individuals at the graduate level, increasing the overall quality and productivity of the institution's academic environment. Enrolling talented foreign graduate students also enhances the reputation of universities as elite global institutions. ICT companies, meanwhile, also have a stake in liberal policies toward foreign students, albeit one that is more indirect. This is not merely because such firms have a stake in the success of national universities, given the role of these universities in the national innovation system. It is also because foreign graduates in technical fields represent a pool of skilled labor that companies can access in the future. Although firms may hire foreign graduates from abroad, foreign graduates from domestic universities are particularly desirable to employers. Such graduates are more readily located, have more familiar credentials, and are more culturally acclimatized than foreign graduates without comparable experience.[68]

Let us now consider the HTC's stake in policies that support global R&D. Both large ICT firms and research universities engage in cross-border R&D collaboration, as I noted in chapter 1. Such firms and universities thus naturally have an interest in permissive policies in this regard, since such openness amplifies the opportunities available for collaborative work. With regard to overseas R&D investments, large ICT firms are likely to be the most interested parties.[69] This interest is multifaceted. First, and most simply, foreign R&D labor may be cheaper than domestic labor. Although this potential motivation is the simplest and most controversial, it does not appear to be the dominant consideration in most cases: Most overseas R&D is still performed by firms from developed countries in other developed countries.[70] Second, foreign R&D centers allow firms to engage in asset-augmenting R&D—to access specific skills or knowledge that complement those available domestically. This motivation has apparently become more important as the increasing pace and complexity of innovation in recent years has increased the desire for expertise from multiple locations.[71] Third, because they are located in foreign markets, foreign R&D centers are well positioned to understand consumers in those markets and to adapt products to local preferences and requirements. That is, they are ideally suited to conduct asset-exploiting R&D. These latter two considerations are also important in motivating R&D alliances with foreign partners, since these partners may also possess specialized knowledge or a better understanding of foreign markets.

Although the HTC is clearly a diverse community, we can expect it to take a strong interest in liberal policies toward global innovation, including both the migration of human capital and global R&D. There are some caveats, to be sure. The academic wing can be assumed to take a stronger interest in policies toward foreign students, although the corporate wing also has a stake in such policies. And while both wings have an interest in global R&D collaboration, the corporate wing (particularly large multinational firms) has a stake in policies toward overseas investments that academia does not. With these caveats in mind, let us turn to our next question: When will the HTC succeed in pursuing these interests?

CONTESTED OPENNESS

In the past, scholars have often focused on business interests as playing a decisive role in shaping national preferences in foreign economic policy.[72] The HTC—consisting not only of leading ICT firms but also prominent research universities—clearly has the potential to be a formidable force in this realm. Most obviously, high-tech companies are highly profitable. As of 2015, the twenty-five most profitable Fortune 500 companies included a virtual "who's who" of the top ICT firms in the United States: Apple, Microsoft, Google, IBM, Intel, Oracle, Qualcomm, and Cisco.[73] Such profitability provides ample resources for lobbying and contributions to political campaigns. Moreover, the positive spillovers of high-tech innovation are substantial, and no policymaker wishes to be blamed for "killing the goose that laid the golden egg."[74] In fact, as the U.S. military has come to rely on commercial products, some government officials now consider the profits of ICT firms a matter of national security: They are needed to drive the R&D

that will help create the next generation of military technology.[75] Although research universities cannot rival the financial resources of leading ICT firms, as a group they possess enormous prestige, personal connections with prominent officials, and impressive geographic diversity. Last, high-tech interests are potentially powerful because of their capacity to inform policymakers about the impact of proposed policy changes.[76] Given the highly technical nature of high-tech research, as well as the uncertainties inherent in policymaking, it is difficult for policymakers to know in advance how policy changes will affect the high-tech sector. ICT companies and research universities are uniquely well positioned to address such questions.

When the HTC presses for openness toward global innovation, it is easy to imagine it will have its way. Yet as the scholarship on interest groups from the past two decades makes clear, powerful business interests do not always succeed in shaping national policy.[77] Although businesses often possess considerable financial resources, such resources hardly guarantee the ability to dictate terms to policymakers.[78] In fact, even when business interests are unified, the business community can struggle to accomplish its political goals.[79] The point here is not that money is unimportant but that it is not the only variable shaping the success or failure of lobbying efforts.[80] The context in which the HTC pursues its interests matters considerably.

The nature of organized opposition is likely to be a critical variable in this regard. Scholars have argued for decades that the presence or absence of opposition plays a powerful role in shaping whether a particular interest group succeeds in achieving its goals.[81] Even so, the potential importance of opposing groups has not always been appreciated: Quantitative studies of lobbying influence have sometimes ignored the activity of rival players.[82] The most likely source of such opposition, in turn, is other organized interests. Although government officials (whether from the legislative or executive branch) can be powerful opponents, relatively comprehensive research indicates that it is more common for interest groups to be opposed by other interest groups.[83] In one recent study, in fact, lobbyists cited other organized interests as the primary obstacle to achieving their goals.[84]

In this study, I am not interested simply in the presence or absence of opposing interests, but in three different types of organized opposition.

First, and most simply, the HTC may face little or no organized resistance, in which case it should be generally successful in achieving its aims. It may be, for example, that other groups involved do not perceive openness in a given area of policy as a threat. Indeed, globalization often presents labor with complex questions, making the specific way in which interests are perceived an important variable.[85] The impact of immigration on wages, for example, is a complicated subject, and the way in which the influx of labor from abroad affects employment and wages remains unclear.[86] Moreover, surveys indicate that American voters welcome high-skilled immigrants far more readily than low-skilled immigrants.[87] With regard to foreign investment, labor's interests may be unclear as well. If foreign investment is viewed simply as an outflow of jobs from the dominant state, high-tech labor is likely to be opposed to it. Alternatively, the work done abroad may be seen as complementary to that done in the dominant state—and that may in fact be the case.[88] Overseas R&D may primarily entail local adaptation to promote sales, for example. In this case, it may be viewed as essential to promoting the dominant state's products overseas and thus very much in labor's interest.

The second possibility is that the HTC will face opposition from labor. The most likely source of opposition is high-tech labor; that is, skilled labor in the leading sector. Recent work on the politics of skilled immigration, for example, has shown that native workers in the same sector can perceive skilled immigrants as a threat.[89] Such perceptions can impede flows of foreign labor when skilled work is subject to licensure requirements and skilled natives gain control over the licensing process at the subnational level.[90] Although skilled natives face more daunting collective-action problems at the national level, they may attempt to shape national policy through national professional associations or by working with the national labor movement.

If this is the case, the question becomes the relative political strength of the HTC and its labor opponents. In general, high-tech labor is likely to have difficulty competing with the HTC for influence. Although financial resources do not determine lobbying outcomes, they can be important when the level of resources is unbalanced between opposing sides, as they are likely to be here.[91] The question then becomes whether high-tech labor can compensate for this weakness by adopting "outside" or "grassroots" tactics, such as pressuring policymakers or employers through mass mobilization.[92] This is likely to be difficult. Labor's ability to mobilize in a given sector is powerfully shaped by the density of union membership among workers in that sector.[93] Unionization rates tend to be low for high-tech workers, particularly in software, who identify more as independent professionals than as a collective.[94] Although high-tech professional associations may ally with the national labor movement, such collaboration faces constraints. Mass mobilization is a complex and costly undertaking, and it is likely to be difficult for high-tech professionals (particularly when many are not union members) to convince the wider labor movement to invest its full capacity in their cause.[95] In addition, the national labor movement may face conflicting incentives on some of the issues of interest here. Specifically, unions confront a dilemma in immigration politics: whether to advocate restrictive policies that close labor markets or to support openness in an effort to bring immigrant workers into the labor movement and maintain standards for all.[96] As a result, the HTC is likely to be relatively successful when facing opposition from labor, if not as successful as when it faces no organized resistance.

The third possibility is that the HTC will face resistance from large citizen groups. These are lobbying groups that mobilize members, donors, or activists around issues other than their vocation or profession, such as the Sierra Club or the National Rifle Association.[97] Such groups can be formidable opponents. In fact, recent research has shown that citizen groups are far more successful at opposing business interests than are labor unions.[98] In this study, large citizen groups have the potential to be effective when they are able to counter the HTC's wealth, prestige, and expertise with their capacity for mass mobilization. Such mobilization can take many forms, ranging from low-effort tactics (e.g., phoning or emailing an elected representative's office) to high-effort tactics (e.g., meeting with a legislator in person). Citizen groups often excel at facilitating the low-effort variety. In particular, recent research indicates that joining a citizen group in the United States greatly increases the likelihood that an individual will contact

Congress, whereas joining a trade or professional group has no such effect.[99] On a large scale, these contacts matter. They inform policymakers not only about the extent of voter sentiment on a given issue, but also about the intensity of that sentiment—information policymakers have trouble acquiring otherwise.[100] More pointedly, a well-coordinated mobilization campaign can demonstrate the ability of a particular group to mobilize voters on election day.[101] As a result, the HTC is more likely to struggle to achieve its goals when it faces opposition from large citizen groups than when it faces either no resistance or resistance from labor.

Taking the preceding discussion as a whole, we can generate the following three propositions regarding the dominant state's level of openness to collaboration. First, when the HTC faces little or no resistance in its push for openness, it will have relatively little difficulty achieving its goals, and the dominant state will adopt a relatively open approach. Second, when the HTC faces resistance from organized labor, policymaking will be more contentious, and policy will be less open than in the first case. As argued, however, labor's ability to compete with the HTC is limited, so restrictions on openness will also be limited. Third, when the HTC faces organized resistance from one or more large citizen groups adept at grass-roots mobilization, it will have greater difficulty in its push for openness, and the dominant state's policy will be less open than in the preceding two cases.

Before we proceed, a word about the scope of the theory I have outlined is in order. The preceding propositions have focused on the relative influence of competing societal interests within the dominant state. By privileging the role of the HTC and its potential opponents in shaping national preferences, I have adopted a liberal theoretical approach, in particular one that combines commercial and ideational liberalism.[102] The theory is not designed, however, to predict policy outcomes when the primary political dynamic involves bargaining between the HTC and the national executive. Focusing on one political process for reasons of tractability and parsimony, as I have done here, is common in the study of foreign economic policy.[103] In some cases, however, state–society bargaining will be the primary dynamic, particularly when the HTC's pursuit of openness conflicts with the executive's preferred means of ensuring national security. Under such conditions, a range of outcomes are possible. The HTC naturally cannot dictate terms to executive agencies on matters of national security. But the executive must be concerned with the fortunes of important national firms and universities. The executive also needs to ensure that restrictions on commerce do not have unintended consequences, particularly when the subject matter is highly technical and the government's information is limited. I will return to this point in the conclusion.

ALTERNATIVE EXPLANATIONS

The theory I have outlined here focuses on the role of high-tech actors and other domestic interests in driving the dominant state's approach to global innovation. Yet there are a number of other possible explanations for the dominant state's policies. Let us now explore several specific alternatives, beginning with a more strategic approach and then turning to other explanations focused on domestic politics.

The most obvious alternative perspective is that the dominant state's approach is state driven and deliberate, rather than society driven and seemingly haphazard. More specifically, the dominant state's policies could reflect concerns surrounding the distribution of gains from global innovation. Prominent realist scholars have argued that the dominant state should be highly sensitive to relative gains when engaging in commerce with rising powers, since the progress of such powers threatens to erode its primacy.[104] Collaboration in innovation would seem a particularly sensitive issue in this regard, given the way in which technological leadership has historically generated and sustained the position of the dominant state.[105] Nonetheless, security concerns need not imply a blanket ban on collaboration between dominant and rising states. Instead, the dominant state may restrict collaboration only when doing so promises to improve its relative position. Other scholars suggest this is likely to be the case when two conditions are met.[106] First, restricting collaboration effectively must be feasible. Unilateral restrictions may be ineffective if the dominant state does not possess a monopoly on the relevant resource or opportunities. Multilateral measures, in turn, may be difficult to arrange. Other countries may not share the dominant state's concerns, and even if they do share its concerns, they may seek to free-ride on its efforts.[107] Second, the restrictions on commerce must be less costly for the dominant state than for the rival power in question. In short, we should expect the dominant state's openness to reflect a calculating approach to relative gains vis-à-vis rising states. I call this the "strategic hegemon" hypothesis.

There are also several alternative explanations preoccupied with domestic politics, but which employ a different focus. The first of these focuses on wider beliefs and attitudes among the general public. Some scholars have argued that issues on which business interests are unified tend to be "ideological, partisan, and highly salient among the public."[108] In these situations, business interests have difficulty prevailing unless they enjoy substantial support from public opinion. It could be the case, then, that the HTC will be most successful when public opinion is favorable to its goals and that it will have the most difficulty when public opinion is unfavorable. I call this the "public opinion" hypothesis.

The second alternative explanation focused on domestic politics concerns political parties and their relationships with key interest groups. Both progressive and conservative parties can be expected to cultivate relations with high-tech interests, given the latter's wealth, prominence, and national importance. Yet such parties are likely to have contrasting relations with the HTC's potential opponents. Labor movements typically have closer ties to progressive parties than their conservative counterparts. Citizen groups, meanwhile, can often be classified as closer to progressive or conservative parties based on the nature of their mission and their membership base. As a result, the power of these potential opponents may be highly contingent, rather than consistent, over time. More specifically, we can predict that organized labor will have substantially greater capacity to resist the HTC when a progressive party predominates, and we can also predict that the ability of a given citizen group to resist the HTC will vary depending on the political fortunes of its preferred party. In short, we should observe wide variations in the ability of labor and citizen groups to counter the HTC's agenda, rather than consistently greater capability on the part of citizen groups, as theorized earlier. I call this the "preferred party" hypothesis.

The third alternative explanation focused on domestic politics concerns whether the HTC is trying to defend or change the status quo. Changing the status quo is normally more difficult than defending it: Policy change requires surmounting all of the relevant veto points in the political system, whereas defending the status quo merely requires success at one veto point. The status quo bias can be reinforced by partisanship, the limited attention of policymakers, fear of unintended consequences, and special interests that mobilize when threatening changes loom on the horizon.[109] We can therefore explain variation in policy outcomes by focusing on the nature of the HTC's challenge. Simply put, the HTC should be relatively successful when it is defending the status quo, and it should have more difficulty when it is trying to change it. I call this the "status quo" hypothesis.

A final alternative explanation focuses on coalition strength. Although coalitions of interest groups are quite popular, studies have often failed to find these alliances effective at shaping policy outcomes. However, recent work has shown that larger and more cohesive coalitions have greater impact on policy, other things being equal.[110] This implies that the HTC will be most successful when both its corporate and academic wings are actively engaged on a given issue than in cases in which one is largely inactive, either because it is free-riding or because its interest in the issue is less strong. In addition, research has found that the average business association is not particularly powerful but that business groups tend to prevail because they are relatively numerous and their power is aggregated when they are unified.[111] This finding implies that the HTC will be most likely to succeed when its corporate wing allies with other business groups in pursuit of a shared goal. Thus, we can make two predictions based on coalition strength. The first is that the HTC will be most successful when both of its wings are actively engaged in a given case. The second is that the HTC will be most successful when its corporate wing allies with other business interests. I call these the "coalition strength" hypotheses.

TESTING THE THEORY

In this book, I test the propositions set forth in this chapter through a series of case studies focusing on different aspects of U.S. policy toward global innovation. The first focuses on the flow of skilled high-tech workers to the United States. The second focuses on the flow of academic labor, particularly students, to U.S. universities. The third focuses on U.S. policy toward global R&D, particularly in the realm of outward foreign investment. These three cases involve varying political dynamics, allowing us to examine the effects of varying levels of organized resistance to the HTC. In addition, the first case study involves varying resistance to the HTC over time, allowing us to make inferences about the effect of this variable while controlling for policy domain.

Each of the next three chapters begins with an overview of U.S. policy in the domain in question. Each then delves into the politics behind the U.S. policymaking process, with particular attention paid to the role of interest groups in shaping U.S. policy. These in-depth case studies perform three key analytical tasks. First, they document the nature of opposition faced by the HTC. As noted, I conceive of three broad types of organized opposition: no resistance, resistance from organized labor, and resistance from citizen groups. Second, each case study documents the interest of the HTC in openness and assesses how successful it was in pursuing this interest in the case in question. This part of the analysis is relatively lengthy and involves detailed process tracing, allowing us to determine the relative importance of the opposition to the HTC in shaping the outcome.[112] Third, the conclusion of each chapter considers how well competing explanations can explain the outcome observed in that case.

Before we proceed, a bit more discussion of measurement is in order. In the preceding paragraph, I noted that each case study will evaluate how successful the HTC is in pursuing its interest in openness. We therefore need an effective means of comparing outcomes across the case studies. Comparing openness across immigration policies is relatively straightforward: The key variable is whether flows are limited by an annual ceiling and, if so, what the ceiling is.[113] It is less obvious, however, how we should compare levels of openness between immigration policies and policies toward global R&D. Global R&D is inherently more difficult to regulate than legal flows of people across national borders, and the two areas involve different metrics as well. Nonetheless, as will become apparent in the following chapters, U.S. policymakers have had opportunities to restrict openness in each of the policy arenas of interest in this book. Under these circumstances, it is useful to focus on the process through which openness is or is not achieved. That is, it is useful to ask to what degree the HTC succeeded in achieving its specific objectives in the case in question. Was it entirely successful, partly successful, or unsuccessful? By focusing on the HTC's level of success, we can compare results across cases that involve different types of economic activity.

In the book's conclusion, the within-case analysis in each of the empirical chapters is complemented by a broader across-case analysis. In this discussion, I also consider how well the alternative explanations outlined in this chapter fare across all three case studies.

SUMMARY

In this chapter, I have developed a theory to explain the dominant state's approach to global innovation. The starting point for this theory is the dominant state's economy and the societal interests that reside in it. Historically, the dominant state has possessed an unusually innovative economy, and this played an important role in generating and sustaining its position in the international system. In recent decades, the actors responsible for innovation in the dominant state have emerged as a potent political force—the high-tech community—with a collective interest in openness to global innovation. The HTC's success in pursuing this goal, in turn, depends on the nature of resistance it faces from other organized groups. In the chapters that follow, I put this theory to the test.

3

THE SWINGING DOOR

Skilled Workers

ALTHOUGH THE UNITED STATES leads the world in innovation, many of its innovators are born abroad. In fact, the number of foreign-born innovators in the United States has increased markedly in recent decades. In 1993, for example, 27 percent of the country's S&E workers with doctoral degrees were born overseas. By 2013, that figure had climbed to 42 percent.[1] China and India, in turn, have emerged as the preeminent suppliers of this high-skilled labor. In 2013, the two countries accounted for more than 36 percent of the foreign-born population in the United States with a doctoral degree in an S&E discipline.[2] Even so, U.S. openness to skilled immigrants has varied considerably over the past few decades. The United States became markedly more open in the late 1990s and early 2000s but less open thereafter.

In this chapter, I explain the curious evolution of U.S. policy. The analysis considers both employment-based (EB) visas (also known as employment-based green cards) and temporary visas (particularly the H-1B visa program). The primary focus is on the H-1B program, since this program has varied so widely over time. In the first section of the chapter, I provide an overview of U.S. policy toward skilled immigration and how it has evolved since 1990. In the second section, I document and explain the expansion of skilled immigration from 1998 to 2004. During this period, the high-tech community faced resistance mainly from organized labor. This was a lop-sided contest in which the HTC recorded a series of legislative victories. In the third section, I focus on the period from 2005 to 2016, when efforts to raise the H-1B cap became intertwined with broader efforts to liberalize U.S. immigration policy. As a result of this change, the HTC also had to contend with fierce resistance organized by large citizen groups. These groups have effectively stymied the HTC's efforts to expand skilled immigration since 2004.

A DYSFUNCTIONAL SYSTEM

Although the United States is often described as a "nation of immigrants," U.S. openness to well-educated workers from abroad has varied widely over time. Over the past quarter-century, U.S. policy has been defined in many ways by the Immigration Act of 1990. The Act raised the number of EB visas from 54,000 to 140,000 per year, which it distributed across five employment categories, and increased the share going to highly skilled workers and their families.[3] The Act also revised the program for admitting skilled workers into the United States on a temporary basis. Starting in the 1950s, the H-1 visa program had admitted workers "of distinguished merit and ability" who were "coming temporarily to the United States to perform services," and no cap was imposed on such visas.[4] In 1989, the H-1 program was split into the H-1A and H-1B visa programs. The H-1A program was designed for nurses, and the H-1B program was designed for other skilled workers in the H-1 category. The 1990 Act made changes to the H-1B visa program and also created the O visa for individuals of "extraordinary ability" in the sciences, arts, education, business, or athletics.

The changes to the H-1B program in 1990 were far-reaching. The Act eliminated the "distinguished merit and ability" formulation and introduced the concept of workers in "specialty occupations." These were occupations requiring both (1) theoretical and practical application of a body of highly specialized knowledge; and (2) attainment of a bachelor's or higher degree in the specialty (or its equivalent).[5] Employers were required to file a labor condition application to hire H-1B workers but were not required to test the domestic labor market (as is required for EB visas). The new H-1B visa was valid for three years and could be extended for an additional three years. The program was capped at 65,000 visas per year.[6]

Taken together, the 1990 Act's provisions on skilled immigration reflected a deliberate plan. The primary architect of these provisions in the House, Immigration Subcommittee Chair Bruce Morrison, sought to limit the role of guest workers and to emphasize instead immigration through the green-card system. As Morrison later put it, "Our goal was to promote American economic competitiveness by using our greatest economic and civic advantage over the rest of the world—almost unique among the nations of the world, the United States does not merely admit foreigners as workers . . . we welcome high skilled individuals from around the world as new Americans."[7] The H-1B program was thus capped, and the number of EB visas was dramatically expanded, and it was hoped that H-1B workers would quickly graduate to immigrant status.[8]

The plan began to unravel even before the Act was passed, however. To promote use of EB visas, Morrison had sought to replace the program's labor certification requirement—a domestic labor market test—with a fee (as well as a more limited assessment of labor shortages). The rationale was that the certification requirement was largely meaningless, involving legal artistry rather than a serious analysis of the labor market, and that certifying many more EB visas each year would generate disruptive processing delays. In contrast, a fee-based system would move applications more quickly and would ensure that employers imported labor only when a compelling need arose. In the end, the leadership of the House Committee on Ways and Means objected to the fees: It saw these as a "tax" and thus not within the purview of Morrison's subcommittee.[9] The EB visa program thus continued to contain a labor certification requirement after 1990.

As the 1990s proceeded, the U.S. skilled immigration system strayed even further from the vision behind the 1990 Act. The slow-moving EB visa system was unappealing to employers, and the limit of 140,000 was never reached in the latter half of the decade. In contrast, the popular H-1B

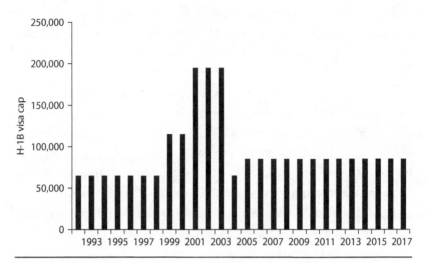

FIGURE 3.1 The H-1B visa cap, 1992 to 2017. *Source*: U.S. Citizenship and Immigration Services, *Characteristics of H-1B Specialty Occupation Workers: Fiscal Year 2015* (Washington, DC: U.S. Department of Homeland Security, 2016), 3–4.

program was expanded in 1998, 2000, and 2004. These expansions included both exemptions for certain classes of applicants and temporary increases in the annual cap, which rose as high as 195,000 from 2001 to 2003 (figure 3.1). A direct result has been large increases in the number of H-1B workers in the United States. In recent years, in fact, some have suggested that the H-1B population has exceeded 650,000.[10] An indirect result has been the deterioration of the EB visa program. With growing numbers of H-1B workers seeking green cards, a time-consuming certification process, and caps on the number of green cards that can be issued for individual countries, an enormous backlog of green card applications has developed, with the waiting time more than ten years in some cases. Since 1990, then, the role of guest workers has expanded, and the transition from guest worker to immigrant status has become a highly problematic process.

For the past two decades, the main users of the H-1B program have been ICT companies, while universities and other research institutions have represented a smaller but steadier source of demand.[11] Demand for the visas among Silicon Valley employers is particularly intense. This was not always the case: In 1995, more than half of H-1B visas went to physical therapists and only a quarter to computer specialists.[12] By 2000, however, 58 percent of all H-1B recipients were in computer-related fields, and 12 percent were in the category of "architecture, engineering, and surveying" (which includes electrical engineers).[13] As the program evolved, India emerged as the primary supplier of these workers, with China in second place. This also was a big change: In 1989, the leading sources of H-1 workers were the Philippines (24 percent) and the United Kingdom (10 percent).[14] Together, China and India accounted for just 3 percent of such workers. By 2000, however, 49 percent of H-1B workers were from India, whereas 9 percent were from China, and 3 percent were from Canada.[15] By 2015, India's share had climbed to 71 percent, and China's share stood at 10 percent.[16]

The increases in the H-1B program—both in terms of exemptions and increases in the cap—came to a halt after 2004, frustrating many high-tech employers and prompting them to look for workarounds. For example, companies may transfer foreign staff from overseas to the United States through the L-1A and L-1B visa programs, which are not subject to an annual cap. When the H-1B cap has posed a problem, therefore, some companies have hired skilled foreign employees overseas and then transferred them to the United States. This workaround is no panacea, however. L-1 applicants must have worked for the company for at least one year before applying, and their applications are subject to different standards and can face considerable scrutiny. Another workaround involves hiring foreign S&E graduates of U.S. universities. In general, foreign graduates are allowed to work in the United States for twelve months under the F-1 visa's Optional Practical Training (OPT) program. In 2008, the Bush administration allowed foreign graduates with degrees in science, technology, engineering, and mathematics (STEM) fields to extend their OPT employment by seventeen months. In March 2016, the Obama administration increased the extension period to twenty-four months.[17] The OPT program is also of limited utility, however, since the program is restricted in duration and scope. Some companies have responded to the H-1B limit by doing more work in overseas locations, but this is not always feasible, and it is more practical for larger firms than for smaller ones.[18] For all these reasons, the HTC has continued to seek increases in the H-1B cap.

Why did the United States expand the H-1B program in 1998, 2000, and 2004? And why has the United States chosen not to expand skilled immigration since then? The remainder of this chapter answers these questions.

AN OPENING DOOR: 1998 TO 2004

In the first half of the 1990s, statutory limits on skilled immigration were not a serious concern for high-tech firms—or for business in general. The increase in the limit on EB visas to 140,000 in the 1990 Immigration Act actually exceeded the business community's request.[19] There were concerns about the limit on the H-1B program, to be sure, but as a prominent business advocate later put it, these were "theoretical" in nature.[20] The visa cap of 65,000 exceeded the business community's requirements when the law was passed, and the cap was never reached in the first half of the 1990s.[21] For its part, the HTC did not play a prominent role in negotiating the limits in the 1990 Act. The "prime movers" in the business coalition lobbying in favor of skilled immigration in 1990 were the U.S. Chamber of Commerce, the National Association of Manufacturers, and the American Immigration Lawyers Association. Although there was some interest from technology firms, "it was not as organized and not as focused as it became in future years."[22]

The HTC became far more interested in limits on skilled immigration—particularly the H-1B program—as the ICT industry boomed in the latter half of the decade. With no cumbersome certification process, the H-1B program became the vehicle of choice for funneling foreign workers to U.S. technology firms. The specific users of the H-1B program varied. Prominent U.S. ICT firms using the program at this time included Cisco, Intel, Microsoft, Motorola, and Oracle, among others.[23] The program also became highly popular with Indian firms, including Infosys, Tata Consultancy Services, and Wipro. These firms developed a business providing H-1B workers as contract labor to U.S. firms—a model that was also adopted by some U.S. firms, including Accenture, Deloitte, and IBM. Universities were less prominent users of the program in the late 1990s, accounting for between 6,000 and 10,000 visas per year.[24] This was partly because the hiring practices of universities were relatively slow, making it hard for them to compete with the private sector for a limited pool of visas. As Kevin Casey, the senior director of federal relations for Harvard University, put it in October 2000, "In the past two years with the accelerated use of the H-1B visas, businesses were burning them up so that by the time we got around to hiring, they were all used up."[25]

With the cap becoming a constraint, prominent technology firms began to lobby the U.S. government to maintain and expand the program. In 1995, the Information Technology Association of America (ITAA) hired Harris Miller, a former congressional staffer and lobbyist with expertise in immigration, as its president.[26] A new lobbying group, American Business for Legal Immigration (ABLI), was set up with financial support from Cisco, Intel, Microsoft, and other ICT firms to focus specifically on the H-1B issue.[27] This shift was part of a broader "coming of age" for many high-tech firms, which had earlier spurned politics in general and Washington in particular.[28] As Miller would recall in 2000, "Five years ago you had almost nobody (who lobbied) . . . they just didn't participate in the political process . . . there was a fundamental belief that if they ignored Washington it would go away."[29] The campaign to raise the H-1B cap would become an example of how much influence high-tech firms could exert collectively.

High-tech interests had a range of advantages that positioned them to succeed. First, by the mid-1990s, lawmakers had discovered that high-tech firms could be lucrative sources of campaign funds. President Bill Clinton tapped into Silicon Valley for his 1996 reelection campaign, even changing his position on securities fraud lawsuits to do so, and the Democratic National Committee benefitted from its largesse as well.[30] Although Clinton would not run again, Vice-President Al Gore was preparing for his own presidential campaign and cultivating support from technology firms. There was also concern among the Democrats that if they spurned Silicon Valley, it would become a base of support for the Republicans.

Indeed, Republican strategists hoped that differences over the H-1B issue would drive a wedge between Silicon Valley and the Democrats, especially Gore, creating an opportunity the GOP could exploit.[31] Last, and just as important, technology firms had tremendous credibility and access to lawmakers. In 1998, one awed journalist wrote of TechNet, a grouping of Silicon Valley executives, "TechNet's ability to gain intimate audiences with officials at the highest levels is truly impressive. Already, (it) has staged over 70 'issue briefings'—small tête-à-têtes frequently likened to 'graduate seminars on the new economy'—between its members and various politicos, from Speaker Newt Gingrich to Vice-President Al Gore."[32]

The main opposition to the effort to increase the H-1B cap would come from labor. In general, high-tech workers tended to avoid labor unions. In fact, less than 2 percent of high-tech workers in North America were union members in the late 1990s, according to the U.S. Department of Labor.[33] Even so, some workers belonged to politically active professional associations. Of these, the Institute of Electrical and Electronics Engineers-USA (IEEE-USA) was the most prominent in opposing H-1B expansion. In March 1998, the IEEE-USA president, John Reinert, told Congress that the "claims of high-tech worker shortages are inflated, the available domestic labor supply is understated, and the wisdom of expanding immigration is overrated" by high-tech companies.[34] Reinert urged Congress to let the labor market correct any existing imbalances rather than resort to immigration as a "quick fix."

The IEEE-USA had roughly 220,000 members as of 1998, but for several reasons it was poorly positioned to compete with high-tech firms in the political arena.[35] First, financial constraints limited what it could spend on lobbyists or campaign contributions. Second, the group was poorly positioned to organize grass-roots protests, since its membership tended to be politically disengaged and libertarian in outlook. The group also suffered from internal divisions, with academic members often unsympathetic to the concerns of those in industry. The IEEE-USA was also under pressure from its umbrella organization, the IEEE, to moderate its position. As a global organization with members throughout the world, the IEEE was concerned that criticism of the H-1B program would vilify some of its non-American members. The IEEE-USA also had trouble rivaling the "star power" and public credibility of those on the other side, such as the CEO of Microsoft, Bill Gates. To be sure, the IEEE-USA could count on some support from other labor groups, particularly the American Federation of Labor and Congress of Industrial Organizations (AFL-CIO), which felt the H-1B program was open to abuse. The influence of the AFL-CIO was constrained, however, by low union density among technology workers. The lack of "worker voices in the advocacy space," as the AFL-CIO's former director of immigration policy put it, made lobbying and grass-roots action on the H-1B issue difficult.[36]

H-1B expansion was also opposed by some anti-immigration groups at this time, but these groups possessed little capacity for grass-roots mobilization. The most established group, the Federation for American Immigration Reform (FAIR), had been founded in 1979. FAIR was small, however: The group claimed to have roughly 70,000 members in March 1998, and others suggest its membership may have been substantially lower.[37] In 1996, a new group—NumbersUSA—was founded with a mission to reduce both legal and illegal immigration. NumbersUSA would eventually eclipse other anti-immigration groups in size and mobilize activists more effectively through the Internet. In the late 1990s, however, NumbersUSA was just finding its feet. It had no full-time lobbying office in Washington at that point, and it had fewer than 4,000 "e-mail activists" as of January 2000.[38] In short, the anti-immigration movement had little ability to compete with the HTC on the H-1B issue in the late 1990s.

It was against this backdrop that high-tech interests would record several legislative victories between 1998 and 2004. The first victory involved playing defense. Restrictions on the H-1B program had been proposed in bills curtailing legal and illegal immigration proposed by the Republicans Lamar Smith and Alan Simpson in the House and Senate, respectively. Businesses and other pro-immigration groups successfully stripped out the restrictions on legal immigration, which ensured that the H-1B provisions would not be passed on a wave of other anti-immigration measures.[39] Even so, the campaign to raise the H-1B cap still faced uncertain prospects as of early 1998. In January, Secretary of Commerce William Daley stated that the Clinton administration did not support an increase in the cap, which he described as "politically not feasible."[40] Instead, he urged technology executives to tap into previously overlooked groups, such as women, minorities, and the disabled, for new sources of workers. He also called for them to work more closely with schools. "Talk to the schools as you would talk to your suppliers," he told business leaders. "Schools supply your most important raw material."[41] The political situation became still more complicated in March, when the General Accounting Office (GAO; now the Government Accountability Office) released a report on the supply of IT workers. In 1997, the Department of Commerce and the ITAA each released a report contending that a shortage of IT workers existed in the United States. In response, the GAO report stated that these reports suffered from "serious" and "major" methodological flaws, respectively, and that more research was needed to ascertain whether a shortage truly existed.[42] This conclusion seemed to support groups resisting H-1B expansion, like the IEEE-USA, which called the shortage claims "inflated."[43]

If high-tech firms were struggling to prove that a labor shortage existed, they would nonetheless prevail in Washington over the next several months. The first victory came in the Senate. Senator Spencer Abraham, who had hired the author of the 1997 ITAA report to become his immigration policy director, sponsored legislation to increase the cap in early March. The Senate approved the bill with relatively little opposition, passing it 78 to 20 on May 18. The effort to raise the cap would face more resistance in the House of Representatives and the White House, however. In the House, Lamar Smith was a key figure as chair of the Committee on the Judiciary, and he was decidedly unenthusiastic about raising the cap. President Clinton, in turn, threatened to veto the legislation if it did not feature substantial protections for U.S. workers. Specifically, the administration wanted the legislation to fund training programs and include safeguards for American workers. The opposition sparked a round of high-tech lobbying. The administration's lead negotiator, Vice-President Al Gore, took calls from officials at companies including Hewlett-Packard and Intel, while a number of CEOs sent letters to the White House.[44]

Ultimately, a solution emerged. Clinton and Smith agreed to substantial increases in the H-1B cap: 115,000 would be allowed in fiscal years 1999 and 2000, and 107,500 would be allowed in 2001. In exchange, funding for worker training and education was provided through the imposition of new fees on visa applications. Worker safeguards, however, were limited to "H-1B-dependent" firms.[45] For firms with more than fifty employees, this would apply only if H-1B workers constituted at least 15 percent of their workforce. This provision ensured that the safeguards would not apply to the most prominent U.S. firms, such as Intel and Microsoft, since these were large enough to employ many H-1B workers without reaching the threshold. After the White House signaled that it would accept these terms, the House passed the revised legislation 288 to 133 on September 24. The new bill nearly stalled when it was reintroduced in the Senate, owing to opposition from Senator Tom Harkin, but high-tech firms persisted. As one report put it, the ITAA and TechNet worked "feverishly behind the scenes" to devise a solution.[46] In the end, one was found: The bill was included in the Omnibus Consolidated and Emergency Appropriations Act for 1999, which was signed by the president on October 21.[47]

The outcome was a resounding victory for the HTC. Whereas in January the Clinton administration had been pessimistic about raising the cap, by July it was merely seeking a compromise solution, and in October, the president signed a bill with safeguards that did not apply to the largest employers. The *Washington Post* reported that "high tech is king of the hill," thanks to the visa bill and other legislative victories.[48] The manager of government affairs for Intel summed it up by saying, "There was a lot of gloom and doom that turned into a double rainbow for us."[49] The vice-president of the ITAA, Jonathan Englund, called it "a banner day," one that marked "a coming of age for tech in wielding influence in the Beltway."[50] President Clinton wasted no time taking advantage of the bill's passage: He went to California for a weekend of fundraising right after it passed the House in September.[51]

Not surprisingly, the labor opponents of the bill were angry. "The White House capitulated to industry groups," said the president of IEEE-USA, John Reinert. "The result is a bill that leaves thousands of U.S. workers at risk."[52] Jack Golodner, the president of the AFL-CIO's Department for Professional Employees, dismissed the safeguards in the bill as "just cosmetic," since they did not apply to the vast majority of companies that used the visa program.[53] One Democratic staffer in the House complained that the administration had sidelined those in Congress who sought stronger worker protections. To be sure, labor did prevail on one front: The increases were temporary, and the cap would revert to 65,000 after 2001. Yet there was more to the story than labor's influence over a Democratic administration. As a former industry representative later noted, lawmakers had a powerful incentive to keep the increase temporary as a means of maximizing campaign contributions over time.[54] In contrast to a permanent increase, a temporary increase ensured that technology companies would continue to solicit key lawmakers' support.

This would indeed be the case: In 2000, high-tech firms made another push to raise the cap. To broaden their support this time, they made an effort to ally with research universities. Higher education lobbyists were initially reluctant to become involved because most of the additional H-1B slots would be taken up by industry, given how hard it was for universities to compete for scarce visas. In response, the high-tech firms offered to modify their proposed legislation. Under the new proposal, H-1B visas for applicants going to work for universities and nonprofit organizations would no longer count against the annual cap, making life easier for both academic and corporate employers. Higher education signed on and agreed to support the bill.[55]

Once again, the primary source of resistance to increasing the H-1B cap came from labor. In February, the AFL-CIO issued a landmark statement that called for legalizing undocumented foreign workers and other pro-immigration measures, but the organization remained opposed to guest-worker programs like the H-1B visa program.[56] Key Democrats in favor of increasing the visas subsequently met with the AFL-CIO president, John Sweeney, in an effort to mitigate his opposition to H-1B expansion, but Sweeney remained negative.[57] The Coalition for Fair Employment in Silicon Valley, which represented many black high-tech workers, expressed concerns about discrimination and also resisted raising the cap.[58] For its part, the IEEE-USA adjusted its stance. The problem, the group argued, was that H-1B workers were like "indentured servants," since they could not leave the employer that had originally sponsored them if they wanted to secure a green card.[59] Rather than simply opposing an increase, therefore, the IEEE-USA argued that H-1B workers should be given expedited EB visas, which would allow workers greater freedom. This proposal stemmed from a genuine belief that such a change would improve matters, but it also reflected a recognition that simply opposing the H-1B program was unlikely to succeed.[60]

The weakness of the opposition quickly became apparent. In May, the Clinton administration showed its support for Silicon Valley by offering to raise the H-1B cap to 200,000 per year over the next three years.[61] The ensuing legislation briefly lost momentum after Democrats in the Congressional Hispanic Caucus sought to expand the bill through amendments known as the "Latino fairness proposals."[62] Among other things, these would have expanded an existing legalization program for asylees, while also allowing undocumented immigrants who entered the United States before 1986 to apply for citizenship. In the end, Democratic leaders decided to pursue these proposals through other means, and a bill focused on the H-1B program was allowed to go the floor.[63]

The bill would pass easily. On October 3, the Senate voted 96 to 1 in favor, and the House passed the bill just hours afterward in a voice vote. The new law raised the cap on H-1B visas to 195,000 for 2000, 2001, and 2003. It also delivered to research universities: Foreign individuals working for higher education, government research, and nonprofit institutions would no longer count against the cap. H-1B workers who had been counted against the cap within the preceding six years were exempted as well. The law also removed per-country limits—mainly a problem for applicants from India—on workers seeking EB visas if additional visas were available in other employment-based categories. The law also created new programs to improve the education and training of U.S. workers in science and technology, and it encouraged the Immigration and Naturalization Service to accelerate the processing of all immigration-related applications.[64]

The new law was a big victory for the HTC, and the role that technology firms played in crafting it was widely noted. According to one legislator, in fact, many lawmakers were simply afraid to defy such powerful interests.

"Once it's clear (the visa bill) is going to get through, everybody signs up so nobody can be in the position of being accused of being against high tech," said Senator Robert Bennett of Utah. "There were, in fact, a whole lot of folks against it, but because they are tapping the high-tech community for campaign contributions, they don't want to admit that in public."[65] On National Public Radio, Lindsay Lowell, a professor at Georgetown University, called the ICT industry "a juggernaut."[66] The weakness of the opposition was equally apparent. The president of the IEEE-USA, Paul Kostek, said that "industry was willing to spend money to vote on this . . . and no matter how eloquently the case [against raising the cap] is presented, it doesn't matter."[67] Or as the journalist Juan Williams said to one opponent, "You guys have no political powers on your side in this fight. Everybody's with the high-tech people."[68] With strong proponents and weak opposition, the outcome was easy to predict.

In the early 2000s, tensions over the H-1B visa issue subsided for a time. This was partly a result of the increased supply of visas, but it also reflected the bursting of the dot-com bubble shortly after the bill was passed, which reduced demand for the visas. In fact, when the H-1B cap reverted to 65,000 in October 2003, the *Wall Street Journal* reported that the cut stirred "relatively little unease" in Silicon Valley because demand for the visas had fallen along with the fortunes of technology companies.[69] By 2004, however, competition for H-1B visas was increasing once again, and the program became the focus of another battle on Capitol Hill.

The focus of contention would become a proposal by Congressman Lamar Smith. Starting in the early 2000s, Smith had softened his stance on immigration and began courting high-tech firms. The shift coincided with the redrawing of his congressional district, which now included the offices of several technology companies.[70] In April 2004, Smith offered a bill that catered to these interests. The bill did not raise the H-1B cap, but it did exempt from the limit up to 20,000 foreign students with a master's degree or higher from a U.S. university—effectively making the limit 85,000.

Once again, high-tech firms and research universities would team up to support the proposal. This time, the collaboration was formalized: In early

2004, ABLI changed its name to "Compete America," and the new group included not only prominent U.S. high-tech firms and industry associations but also higher education interests.[71] The latter included the two most prominent groups representing research universities—the Association of American Universities and the Association of Public and Land-Grant Universities— as well as the National Association for Foreign Student Affairs.[72] Sandra Boyd, the first chair of Compete America, later recalled that there was a deliberate effort to create "an employer-driven coalition" that included both business and higher education groups "because we had issues in common."[73] Showcasing the united front, thirty-one prominent corporate and academic leaders—including the CEO of Microsoft, Steve Ballmer, the president of Oracle, Safra Catz, and the president of MIT, Charles Vest—sent a letter to Congress later that year supporting Smith's proposal for a new exemption.[74]

Once again, the proposal would meet with resistance from both the AFL-CIO and the IEEE-USA.[75] In addition, there was more resistance from the anti-immigration movement at this time.[76] The key group in this regard was NumbersUSA, which launched a grass-roots telephone campaign to defeat the bill.[77] Even so, NumbersUSA was still relatively small, with roughly 16,000 e-mail activists at the start of 2004.[78] Its small size, combined with the limited time it had to rally its members against the bill, limited the amount of pressure it was able to put on Congress.[79] As a result, the HTC would prevail. In December 2004, during Congress's lame-duck session, Smith's proposal was incorporated as an amendment into the omnibus appropriations bill for fiscal year 2005. The passage of the appropriations bill was a priority for both parties, and it passed with bipartisan support.[80]

From 1998 to 2004, the HTC recorded a string of victories over organized labor on the H-1B issue. By itself, labor was no match for the political "juggernaut" that the HTC had become. But by 2004, it was evident that high-tech interests were starting to face resistance from a new kind of opponent: anti-immigration groups specializing in grass-roots mobilization. Growing resistance from these and other citizen groups would stymie the HTC's efforts to raise the H-1B cap after 2004.

FORMIDABLE OPPONENTS: 2005 TO 2016

Following 2004, the politics surrounding efforts to expand the H-1B visa program changed dramatically. Presidents George W. Bush and Barack Obama would launch ambitious efforts to liberalize U.S. immigration policy, targeting not only skilled immigration but also the millions of undocumented immigrants in the country. In this context, the HTC would have to contend with greater resistance than ever before. Comprehensive bills that legalized undocumented immigrants would encounter concerted resistance from citizen groups, particularly NumbersUSA but also (under Obama) conservative ideological groups from the Tea Party movement. Targeted bills expanding skilled immigration, meanwhile, would now meet resistance from advocates of comprehensive reform. Latino groups in particular understood that bills addressing skilled immigration separately would weaken business support for comprehensive reform, making it impossible to pursue the legalization of undocumented immigrants. Faced with these formidable opponents, the HTC would repeatedly fail to achieve its goals.

THE BATTLES OF 2006 AND 2007

Following his reelection in November 2004, President George W. Bush would seek a comprehensive liberalization of U.S. immigration policy, one that would grant legal status to many undocumented immigrants already in the United States. The president had announced his hopes in January 2004, but the initiative was resisted by members of his own party, and the administration made little effort to promote it that year.[81] As late as January 2005, in fact, advocates of legalization lamented Bush's lack of commitment to the cause.[82] As 2005 progressed, however, the administration began pursuing its plan with greater zeal. Although controversial within the GOP, Republican strategists hoped the initiative would expand the party's appeal among the nation's fast-growing Latino population.[83] Not to be outdone, Democratic leaders cast their party as the true champion of comprehensive reform and "smart policies that help secure all Latino families and all

Americans," as Senator John Kerry put it.[84] Going forward, therefore, the focus would be on expanding skilled immigration as part of a broader liberalization package.

For its part, the HTC remained interested in liberalizing policy. Despite the exemptions secured in 2000 and 2004, the H-1B cap for fiscal year 2006 was reached in August 2005. Two months later, Compete America wrote lawmakers that further increases to the cap were needed to ensure American "competitiveness."[85] The group also called for liberalizing the EB visa system, which was now backlogged. Compete America warned that the increasingly dysfunctional EB system was making it "nearly impossible" for skilled workers from China and India to gain permanent residence.[86] Other groups aimed to liberalize policy as well. Broader business support for immigration reform came not only from the U.S. Chamber of Commerce but also from groups with a particular interest in immigrant labor, such as the National Restaurant Association and the National Council of Agricultural Employers.[87] The prospect of legalization for undocumented immigrants was strongly supported by Latino groups, which helped to organize large pro-reform demonstrations around the country in 2006.[88] The Roman Catholic Church, with strong roots in Latino communities, was also supportive. The support for comprehensive liberalization coalesced in pro-immigration umbrella groups like the National Immigration Forum.

The pro-reform effort also benefited from divisions within organized labor. While labor was generally supportive of legalizing undocumented immigrants, it was now divided over guest-worker programs. In 2005, the Service Employees International Union (SEIU) and several other groups broke away from the AFL-CIO to form the "Change to Win" labor coalition. The split was partly a result of differences of opinion over immigration: Whereas the AFL-CIO opposed guest-worker programs, including the H-1B program, the SEIU was open to such programs as part of a broader reform effort that would increase its capacity for organizing workers.[89]

Despite the split within labor, the opposition facing the HTC was more formidable than ever before, thanks to an increasingly vigorous anti-immigration movement. NumbersUSA in particular had tapped into the Internet to expand its membership base and mobilize its activists. With immigration reform an increasingly hot topic after 2004, the group's membership grew rapidly. Whereas the group had 16,000 activists in January 2004, this figure rose to 126,000 in January 2006, 236,000 in January 2007, and then 529,000 in January 2008.[90] The group's email distribution list was far larger. By mid-2007, in fact, the list reportedly included 1.5 million addresses.[91] With this growing membership base, NumbersUSA would become a potent force in immigration politics. The group provided its subscribers with summaries of new bills and notified them of important legislative actions in advance. This would provide opportunities for large numbers of anti-immigration voters to weigh in on legislation in a timely and decisive manner. NumbersUSA also developed a scorecard system showing every vote by every member of Congress on immigration-related issues and assigning a letter grade to each member. For Republicans, a poor grade was fodder for primary challengers. The group's membership was also highly organized and active in visiting members of Congress. In fact, NumbersUSA would tailor office visits to maximize impact: If a particular member was known to be of a certain religious faith, for example, NumbersUSA would try to send members of that denomination to the office to make their case.[92] In short, the anti-immigration movement had become a much tougher opponent.

The 2006 Battle

The first test would come in 2006. Despite their newfound energy, anti-immigration groups would have difficulty prevailing in the Senate. The Republican majority leader, Bill Frist of Tennessee, had announced that he would retire from the Senate and was known to be mulling a bid for the White House in 2008. "Frist is in his last eight months in the Senate," Larry Sabato of the University of Virginia observed in March. "He's much more interested in being a presidential candidate than majority leader at this point."[93] Frist, therefore, could not be challenged in a Senate primary contest. Moreover, he had clearly calculated that passage of a comprehensive immigration bill would help his presidential bid. Frist was thus known to support the Bush administration's push for comprehensive reform. While many Republicans were reluctant to support him, Frist could count on Democratic senators to support a bill that appealed to both Latino groups and high-tech firms. For their part, the anti-immigration groups understood that blocking passage in the Senate would be difficult. "Anything can happen in the Senate," said NumbersUSA's Caroline Espinosa. "In the end it really boils down to what happens in conference."[94]

In mid-March, Frist introduced the Securing America's Borders Act, which aimed to reduce illegal immigration and to increase legal immigration. Frist was unable to secure support to cut off debate on the measure, however, so on April 7, a new bill was introduced. The bill contained a raft of measures designed to appeal to different groups.[95] It devoted funds to increase border security while also creating a pathway for some undocumented immigrants to become citizens. It also dramatically expanded the EB and H-1B visa programs. To reduce the backlog on the former, the bill raised the cap on EB visas to 450,000 until 2016 and then set it at 290,000 in 2017, among other measures. The bill increased the cap on H-1B visas from 65,000 to 115,000, and it authorized an additional 20 percent increase whenever the previous year's limit was met. It also exempted from the limit all foreign students with advanced STEM degrees from U.S. universities. The existing exemption of 20,000 would be applied to graduates of foreign universities. With Frist's support, the Senate passed the bill on May 25, 62 to 36. Most of the votes (38) came from Democrats and one independent, with 23 votes coming from Republicans. Frist praised the bill's passage as "a success for the American people."[96]

Attention now turned to the House. The House had passed its own immigration bill in December 2005, but it was focused on enhancing border security and penalizing those assisting undocumented immigrants. The bill also aimed to change undocumented presence in the United States from a civil offense into a felony. The key question was, would the House leadership now attempt to reconcile this bill with the sweeping reforms passed by the Senate? Speaker Dennis Hastert was the pivotal figure at this moment. In April, Hastert had signaled a willingness to broaden the House's bill in conference with the Senate.[97] In early May, White House officials were quietly optimistic that they could count on the speaker to rally enough Republicans behind the reform effort.[98]

The White House would be disappointed. NumbersUSA launched a targeted lobbying effort aimed at Hastert, one that underlined how his leadership would be damaged if he were to pass immigration reform without support from a majority of the Republican caucus.[99] Hastert was sensitive to such threats. In late 2003, he had articulated the "Hastert rule," which maintained that "the job of speaker is not to expedite legislation that runs counter to the wishes of the majority of his majority."[100] Hastert's statement followed instances in 2002 and 2003 when he had relied heavily on Democratic members to pass legislation on immigration and campaign finance, provoking discontent within his own caucus. Although he did not always follow the rule thereafter, he faced particularly fierce resistance on immigration in May 2006, thanks to the growing anti-immigration movement. The anti-immigration Immigration Reform Caucus (IRC) in Congress had grown from sixteen members when it was founded by Congressman Tom Tancredo in 1999 to ninety-seven members by mid-2006—and almost all of them were Republicans.[101] This growth reflected the joint efforts of Tancredo and NumbersUSA.[102] NumbersUSA had asked their members to contact their representatives to ask them to join the IRC. Members of Congress who did join, in turn, received additional points in the NumbersUSA grading system. The group also targeted specific members viewed as potential candidates for the IRC. Gil Gutknecht of Minnesota, for example, joined the IRC after pressure from constituents working with NumbersUSA. The group also worked closely with the leadership of the IRC to make the caucus effective. In early 2002, in fact, two NumbersUSA lobbyists described themselves as "virtual staffers" in Tancredo's office.[103] In particular, NumbersUSA's Rosemary Jenks, a Harvard Law School graduate, worked "almost on a daily basis" in Tancredo's office, providing legal analysis of pending bills.[104] Between the IRC and other critics of legalization, GOP lawmakers estimated that 75 percent of House Republicans were opposed to even a watered-down version of the Senate bill.[105]

Faced with this level of resistance, Hastert sought shelter behind the rule that bore his name. On May 23, a Hastert spokesperson stated that the speaker would invoke the majority-of-the-majority rule with regard to immigration.[106] One Republican senator called this "a death blow" for immigration reform that year, which indeed it was.[107] Although Congress would pass additional legislation on border security in the run-up to the elections, Hastert had killed comprehensive reform for the year—and with it any hope of raising the H-1B cap.

Shortly thereafter, pro-immigration advocates conceded that they had been outfought by the anti-immigration movement in 2006. "The restrictionists are very loud," said Angela Kelley, the deputy director of the National Immigration Forum. "I don't think they're very large, but they're very loud, and they're able to focus their numbers in a really powerful way by having lots of member contacts. And we don't do that as well on our side."[108]

The 2007 Battle

The midterm elections in November 2006 offered a new opportunity to pass a comprehensive bill. The Democratic Party took control of both Houses of Congress, and President Bush remained committed to the reform effort. The HTC remained committed to the task as well. In April, a spokesperson for Compete America commented with regard to raising the H-1B cap, "Right now we're focused on a comprehensive (immigration bill) being the vehicle."[109] On May 9, momentum began to build when the Senate majority leader, Harry Reid, introduced the Secure Borders, Economic Opportunity and Immigration Reform Act of 2007.

As the 2006 bill had, the new bill strove to increase border security while also creating a pathway to citizenship for some undocumented immigrants. And like the 2006 bill, the new legislation raised the H-1B cap to 115,000, with the limit potentially rising depending on demand, and it exempted from the limit all foreign students with advanced STEM degrees from U.S. universities. While high-tech firms welcomed these latter changes, companies were upset with the increased fees and new regulations associated with the H-1B program.[110] The proposed changes to the EB visa system were also controversial. Although the bill aimed to reduce the EB visa backlog, it replaced the existing employer-based system with a merit-based points system that admitted workers on the basis of skills, education, English proficiency, and other factors—a move critics likened to Soviet-style central planning.[111]

If high-tech interests were somewhat concerned about the bill, other groups were strongly opposed. The AFL-CIO was particularly concerned about provisions creating a new guest-worker program to fill mainly low-skill jobs, with 400,000 visas initially with potentially more in later years.[112] Labor advocates also had concerns about the legalization measures, which they saw as excessively onerous. Anti-immigration groups, meanwhile, decried the legalization of undocumented immigrants as "amnesty."[113] They also mobilized opinion against the bill through media outreach. As one Democratic staffer who worked on the issue later put it, "NumbersUSA and FAIR managed to convince Fox News back then to be their twenty-four-hour news channel of the anti-immigrant point of view."[114]

In early June, Reid attempted to end debate on the bill. The effort was derailed when the Democratic senator Byron Dorgan, supported by labor, succeeded in adding a sunset clause to the new guest-worker program. The "poison pill" amendment passed 49 to 48 with votes from Democrats and four Republicans. The four Republicans hoped to kill the bill (or were under pressure to kill it) because of the "amnesty" provisions.[115] Subsequent cloture votes on the bill failed by wide margins with little Republican support. Following lobbying from the White House, the Senate would take up the matter again in late June. The bill was reintroduced with a few modifications, including more funding for border security to entice wavering Republicans.[116] Reid also opened the door to a limited number of amendments. Technology companies worked hard to seize the opportunity. As the *New York Times* reported, the Microsoft executives Bill Gates and Steve Ballmer were leading "a small army of high-tech executives to Capitol Hill" in an effort to shape the evolving legislation.[117] More generally, the business community hoped the bill would pass so it could be modified to address their concerns (regarding the points system for EB visas, for example) at the conference stage.[118]

Both labor and the anti-immigration movement resisted this last push, however. NumbersUSA was particularly active. The president of

NumbersUSA, Roy Beck, would claim that his group flooded Congress with more than 2 million faxes arguing against the immigration overhaul in May and June.[119] On June 28—the day that the Senate was scheduled to vote on cloture—anti-immigration activists inundated Congress with so many angry phone calls that the Capitol switchboard was shut down.[120] The intensity of the opposition became a concern not only for Republican senators but also for some centrist Democrats as well.[121] The protests prompted the Republican senator John Ensign of Nevada to remark, "The intensity level and the passions on this bill, we've never seen anything like it. Not even close."[122]

NumbersUSA also targeted key senators. The Senate minority leader, Mitch McConnell, and the Senate minority whip, Trent Lott—prominent Republicans in favor of legalization—were particular targets. McConnell, in fact, had been integral to bringing a bill back to the Senate floor.[123] In response, NumbersUSA mobilized grass-roots efforts against the bill in McConnell's and Lott's home states of Kentucky and Mississippi— both of which were home to sizable numbers of NumbersUSA activists.[124] The effort included protests, radio coverage, and television advertisements. Lott, for one, was impressed. "Those really pushing for the bill have not been as effective as those pushing against it," the senator said in late June, on a day when NumbersUSA surrounded his office in Jackson, Mississippi, with petitioners and inundated it with phone calls.[125] Even so, Lott was a hard target. He was evidently not planning to run for reelection: He had toyed with retiring before his 2006 reelection, and he would actually retire from the Senate in late 2007.[126] Despite NumbersUSA's pressure, therefore, Lott would vote in favor of cloture on June 28.

McConnell was another matter: The senator would seek reelection in 2008. In response to the grass-roots pressure, McConnell virtually disappeared from view in the days leading up to the vote.[127] In the end, he would show up on June 28 to vote against the bill—a remarkable turnaround for the president's ally. McConnell subsequently explained his decision on the Senate floor. He praised the bill, which he said "represented the best chance we had of getting to our goal." Yet the resistance within his home state was too much to ignore. "I heard from a lot of Kentuckians. Thousands of smart, well-informed people called my offices to talk about this bill. They did not like [this bill]. . . . And to every one of them, I say today, 'Your voice was heard.'"[128] The president of NumbersUSA, Roy Beck, later recalled hoping to stop the bill even if McConnell had voted for it, but the minority leader's about-face made its defeat a certainty.[129] In the end, the cloture motion gained only 46 votes, with 53 against. For the second year in a row, the effort to pass a comprehensive bill had failed.

Not surprisingly, the coalition in favor of reform was displeased with the outcome. The president of the ITAA, Phil Bond, said he was "tremendously disappointed."[130] Yet the pro-reform camp also recognized that they had been outfought once again—by NumbersUSA in particular. "NumbersUSA initiated and turbocharged the populist revolt against the immigration reform package," said Frank Sharry, the executive director of the National Immigration Forum.[131] On another occasion, Sharry said, "You have to give them credit: the phone calls, the faxes, the people who show up at town halls and meetings—you have to say NumbersUSA is behind a fair amount of that."[132] And years later, Sharry would recall, "They generated a huge volume of opposition to the bill, and it was a big factor in our defeat."[133] The HTC and its allies were well aware of who had defeated them, and how they had done it.

THE OBAMA YEARS

The election of Barack Obama as president in November 2008 opened a new chapter in U.S. immigration politics. As a candidate, Obama had offered some support for raising the H-1B cap, but he was more interested in comprehensive reform.[134] This presented the HTC with a dilemma. Although publicly in favor of comprehensive liberalization, technology firms seemed reluctant to invest in an initiative that was so uncertain. "It is not clear how much the [tech lobbyists are] investing," said a congressional staffer in 2010. "We'd like to see them step up more."[135] High-tech's hesitation had a rationale: Having failed in pursuit of comprehensive legislation in 2006 and 2007, technology firms were skeptical that a greater investment would be worthwhile at this time. "The calculation has to be made about the realistic chance of passage. . . . You're not going to call in a CEO if the bill won't pass," as one executive said.[136]

Focusing just on skilled immigration, however, was difficult. With a comprehensive immigration bill being mooted, a bill targeting skilled immigration now faced quiet opposition from other pro-liberalization groups. As Dean Garfield, the president of the Information Technology Industry Council stated in April 2010, "You would have people [in the pro-immigration coalition] who would argue against our initiative" if the high-tech sector tried to pass a targeted bill on skilled immigration.[137] Or as a higher education lobbyist later put it, "The people who would be left off the train don't want the train to go forward without them, so they will stop it."[138] Latino groups in particular had much to lose from the passage of separate legislation on high-skill immigration, since such legislation would reduce the incentive for high-tech firms to support a bill legalizing undocumented immigrants.[139] And Latino groups were formidable. The largest one, the National Council of La Raza, was a Washington-based umbrella organization with nearly 300 community-based affiliates nationwide servicing roughly 5 million Latinos around the country.[140] Latino concerns were particularly important for Democratic lawmakers, who were not eager to flout a group that had strongly supported the party in the 2008 elections.[141]

Faced with this dilemma, the HTC would pursue both targeted and comprehensive reform efforts during Obama's tenure in office. Both approaches would prove unsuccessful.

A Narrow Bill?

Following the midterm elections in 2010, the GOP retook control of the House but not the Senate. During the campaign, Republican lawmakers had said they would prioritize skilled-immigration measures.[142] An initial attempt focused on the backlog for EB visas. In September 2011, Republican Congressman Jason Chaffetz introduced the Fairness for High-Skilled Immigrants Act. The bill eliminated per-country limits on EB visas, helping applicants from China and India in particular, while raising per-country limits on family visas to 15 percent. Compete America and other high-tech groups supported the bill, whereas NumbersUSA took no stance on the bill since it resulted in no net increase in immigration.[143] The bill passed the House easily in November, 389 to 15.

Nonetheless, the bill would founder in the Senate. On the Republican side, Senators Charles Grassley and Jeff Sessions both opposed the bill. Grassley would relent in mid-2012, after provisions were added regarding the prevention of H-1B visa fraud and abuse, but Sessions would not.[144] There were also concerns on the Democratic side. Senator Robert Menendez, a prominent member of the Congressional Hispanic Caucus, wanted additional measures to facilitate family reunification.[145] Including such measures would have made the bill even more difficult to sell on the Republican side, however. The bill thus died in the Senate.

While this was happening, a second initiative focused on EB visas was also underway. In late 2011, negotiations began between the chair of the House Judiciary Committee, Lamar Smith, and the chair of the Senate Judiciary Subcommittee on Immigration, Refugees, and Border Security, Charles Schumer.[146] As the discussions proceeded in 2012, the focus was on making 55,000 EB visas available to foreign-born graduates of U.S. universities with degrees in STEM disciplines and eliminating the diversity visa program, which provided a comparable number of visas to immigrants from countries underrepresented in the United States. The emerging bill was naturally attractive to universities and high-tech firms. The IEEE-USA was also supportive, since the bill represented a shift away from H-1B visas and toward EB visas.[147] NumbersUSA, which was consulted by Smith, was willing to stay on the sidelines if there would be no net increase in immigration.[148]

The search for a compromise hit a snag, however. To placate Latino groups, Schumer wanted to include changes that would facilitate family reunification for foreigners in the United States with relatives abroad. According to Schumer's staff, Smith agreed to small changes in this regard, but he was unwilling to go as far as Schumer wanted.[149] The sparring became public in September, when Schumer and Smith introduced competing bills. Smith's bill, the STEM Jobs Act of 2012, eliminated the diversity visa program in favor of STEM visas. Schumer's bill, in contrast, created a two-year pilot program for 55,000 STEM visas without eliminating the diversity visa program.[150] Whereas Schumer's bill was not put to a vote, Smith's was. On September 20, the House voted on the bill under suspension rules, a fast-track procedure that limits debate but requires a two-thirds majority to pass. Democrats charged that Smith's attempt to pass the bill with Democratic votes was designed to pressure Schumer over the final content of the bill, but the legislation failed to acquire the supermajority needed.[151]

Smith's bill was reconsidered following the November elections. The concerns of Latino groups loomed even larger at this point, however, since they had helped propel Obama to reelection. In late November, the White House said that it "strongly supports legislation to attract and retain foreign students who graduate with advanced STEM degrees."[152] Nonetheless, it continued, it could not support "narrowly tailored proposals that do not meet the president's long-term objectives with respect to comprehensive immigration reform." The statement noted that a more comprehensive effort should also address "important priorities such as establishing a pathway for undocumented individuals to earn their citizenship"—a clear nod to Latinos. Days after Obama's statement, the bill passed the House, with most Republicans and twenty-seven Democrats voting in favor.[153] With the White House uninterested, the Democratic leadership in the Senate chose not to take up the bill, ending its chances of becoming law. In March 2013, Schumer explained to technology lobbyists why the Democrats could not support such legislation: "If there's an attempt to just try and pass high-end, high-tech immigration, guess who will be furious? The Hispanic community."[154]

A New Push for Comprehensive Reform

New hope for a comprehensive bill was born just after the STEM Jobs Act died. In January 2013, President Obama made it clear that comprehensive immigration reform would be a top priority during his second term. That same month, the bipartisan "Gang of Eight" in the Senate unveiled a framework for a comprehensive bill. One member, Senator Schumer, asked business and labor to come up with language both could support, which led to a joint statement of principles in February.[155] Proponents were also more media savvy than they had been in the past. Led by Marco Rubio, Gang of Eight senators courted the conservative media in an effort to garner support for a comprehensive overhaul. The effort did not win over the radio host Rush Limbaugh, but it was relatively successful with Fox News.[156]

The HTC invested heavily to secure a bill they could support. Led by Microsoft, leading U.S. technology firms dominated the list of top lobbyists on immigration in 2013 (table 3.1). Microsoft's overall lobbying expenses jumped from $8.1 million in 2012 to $10.5 million in 2013 as its activity on

TABLE 3.1 TOP LOBBYING CLIENTS ON IMMIGRATION, 2013

CLIENT	NUMBER OF REPORTS FILED
1. Microsoft	64
2. Intel	22
3. Consumer Electronics Association	20
4. Facebook	19
5. Oracle	18
6. Business Roundtable	17
7. Bipartisan Policy Center	16
8. Qualcomm	16
9. National Association of Home Builders	15
10. U.S. Chamber of Commerce	15
11. Cognizant Technology Solutions	14
12. Computer Sciences Corp.	14
13. Entertainment Software Association	14
14. Google	14
15. Motorola Solutions	14

Note: Figures are based on the number of reports filed with the Senate Office of Public Records.
Source: Center for Responsive Politics, "Annual Number of Clients Lobbying on Immigration," 2017, *Opensecrets.org*, www.opensecrets.org/lobby/issuesum.php?id=IMM.

this issue increased.[157] In fact, U.S. high-tech firms' access to lawmakers in the first half of 2013 was described as "extraordinary."[158] High-tech firms were so deeply involved in drafting the legislation that their lobbyists sometimes learned the details of new provisions before some senators in the Gang of Eight did. In addition, in April 2013, the CEO of Facebook, Mark Zuckerberg, and other Silicon Valley executives founded FWD.us, a nonprofit group designed to advocate immigration reform though media advocacy and other channels.[159]

The bill that eventually emerged in the Senate, the Border Security, Economic Opportunity, and Immigration Modernization Act of 2013, was a bold attempt to overhaul the U.S. immigration system. The bill made it possible for many undocumented immigrants to gain legal status and eventually citizenship, ensuring support from Latino groups and other progressive organizations. At the same time, it invested $46 billion in border security measures over ten years. The bill also created a new W visa for lower-skilled workers, along with a raft of other measures. Most relevant here, the bill also featured a variety of provisions to expand and reform skilled immigration.[160]

First, the EB visa program saw a substantial expansion. Although the limit of 140,000 visas was maintained, the bill recaptured unused visas from previous years, eliminated per-country limits, and created several important exemptions. These included exemptions for spouses and children of visa recipients, EB-1 visa recipients, and foreign students graduating from U.S. universities with advanced degrees in STEM fields or a doctorate in any field. The bill also created a new pathway for skilled immigration: a merit-based points system that would allocate 120,000 to 250,000 visas per year based on education, experience, and other credentials. Half of these visas were reserved for high-skill immigrants, whereas the other half were reserved for lower-skilled immigrants. In addition, the bill created a new EB visa (the EB-6) and a new non-immigrant visa (the X visa) targeting startup entrepreneurs.[161]

The bill also made changes to the H-1B program. The cap was increased to 115,000 visas per year, with further expansions up to 180,000 possible based on market demand and unemployment data. The original bill contained a range of protections to ensure that domestic workers would not be adversely affected by this expansion—the result of the negotiations between business and labor. In response to pressure from high-tech firms, and over the objections of the AFL-CIO, the final bill eased some of these requirements while applying others only to H-1B-dependent firms.[162] Moreover, the calculation of H-1B dependent was modified so that major U.S. firms near the limit (such as Facebook) could avoid falling into the category, particularly by sponsoring H-1B workers for EB visas.[163] The leading U.S. firms, as labor's lead negotiator later put it, "got their way."[164] The AFL-CIO continued to support passage of the bill after these changes, but some affiliates now opposed it or supported it with less enthusiasm.[165]

Indian firms using the program were dismayed by the bill, since they typically qualified as H-1B dependent. Among other changes, H-1B-dependent firms were prohibited from placing H-1B workers with other companies—a practice widely employed by Indian firms. In addition, while the new three-tier wage structure was designed to increase the wages of all H-1B workers, H-1B-dependent firms were singled out here, as well, and required to pay their H-1B workers at or above the second tier. The bill also capped the combined number of H-1B and L-1 employees at 75 percent of a company's U.S. workforce, a figure that would fall to 50 percent in 2017. Firms in which H-1B and L-1 workers comprised 30 to 50 percent of the workforce, meanwhile, would have to pay a $5,000 filing fee for each new application. Taken together, the provisions would have shifted the low-cost IT outsourcing market away from H-1B-dependent firms, including such Indian firms as Infosys, Tata Consultancy Services, and Wipro, toward U.S. competitors with enough American employees to avoid the thresholds, such as Accenture and IBM.[166] The president of India's main software trade body decried the provisions as "discriminatory," but they remained in the bill.[167]

Indeed, the HTC was strongly supportive of the bill overall and wanted to see it pass. "We are very pleased with the progress and happy with what's in the bill," the director of government relations for Intel said in May.[168] In June, more than one hundred executives from high-tech firms and industry associations signed a letter calling on the Senate to pass "this critically important legislation."[169] Led by the American Council on Education, meanwhile, fourteen higher education groups "strongly endorse[d]" the bill, calling it "an historic opportunity toward creating an immigration system with bipartisan support that better serves the needs of our country."[170] On June 27, the Senate passed the bill 68 to 32, with fifty-two Democrats, fourteen Republicans, and two independents voting in favor.

The overhaul effort would face greater resistance in the Republican-led House, however. The reasons for this resistance were complex. Anti-immigration groups were once again opposed. But they had trouble organizing the same kind of grass-roots resistance they had in 2006 and 2007. Even the president of NumbersUSA conceded that the anti-immigration movement was not as mobilized in 2013 as it had been in 2007.[171] Yet if the anti-immigration groups were less active, another source of resistance was more active: the Tea Party. The Tea Party movement had come of age in the 2010 midterm elections, when it helped Republicans win back control of the House, and it had helped to inspire a more intense brand of fiscal conservatism in the GOP. Although it began as a grass-roots movement, parts of it coalesced into a collection of professional activist organizations, including the Tea Party Patriots and the Tea Party Nation. Several other conservative groups also became closely associated with the movement, including Americans for Prosperity, Club for Growth, FreedomWorks, and Heritage Action for America, an advocacy group associated with the Heritage Foundation. The movement suffered a set-back in the 2012 elections, when President Obama was reelected and many Tea Party–backed Senate challengers were defeated. In the spring of 2013, however, the Internal Revenue Service disclosed it had been giving extra scrutiny to some conservative groups' requests for tax-exempt status, focusing in particular on those with names that included such words as *tea party*, and this reinvigorated the movement.[172] By late 2013, Tea Party groups had a "core membership" of roughly 500,000 individuals and more than 382,000 followers on Twitter.[173]

The Tea Party quickly became an important source of opposition to immigration reform. In the first half of 2013, to be sure, some conservative groups were hesitant to weigh in on the issue. In particular, Americans for

Prosperity, the Club for Growth, and FreedomWorks indicated that they were hesitant to get involved in this particular legislative battle.[174] In addition, a cofounder of the Tea Party Express, Sal Russo, suggested that he saw immigration reform, including a path to citizenship for some undocumented immigrants, as inevitable.[175] More generally, however, opposition from Tea Party groups was described as "intense and vociferous."[176] One of the largest groups, the Tea Party Patriots, was ardently opposed. An umbrella organization with more than 2,000 local chapters as of mid-2014, the group argued that immigration reform, particularly the provision of a pathway to citizenship for undocumented immigrants, would lead to massive increases in federal spending.[177] "How soon after the immigration bill goes into effect will the current population of illegal immigrants receive Obamacare benefits?" the Tea Party Patriots asked in May.[178] The Heritage Foundation also argued that the bill would prove extremely costly, and Heritage Action for America called it "bad news."[179] Another group, the Madison Project, called the bill "comprehensively flawed."[180] Moreover, the opponents were far more energized than those on the sidelines. As Whit Ayres, a Republican pollster put it in July, House members were "reacting to what they hear from people who take the time to write, call and email their offices. There aren't any Republicans that are going to write their congressman demanding a path to citizenship. It's not the kind of thing that makes you charge the ramparts."[181]

The Tea Party opposition was empowered by the organizational groundwork laid by NumbersUSA. Following the 2010 elections, NumbersUSA courted Tea Party groups as allies against immigration reform, even hiring a Tea Party "liaison" in 2011 to work with Tea Party leaders and speak at Tea Party events.[182] Many members of NumbersUSA joined Tea Party organizations as well, so there was significant overlap in the grass-roots base of each movement.[183] This meant that many Tea Party members received information and updates on immigration issues from NumbersUSA. The overlap between the movements was also evident in the House of Representatives. As of March 2011, roughly 70 percent of the House Tea Party Caucus's members were also members of the anti-immigration Immigration Reform Caucus.[184]

The opponents would channel this energy and organization into stopping the House leadership from pursuing immigration reform. Looking for support in the House, pro-immigration groups identified the majority whip, Kevin McCarthy, as a crucial potential supporter in 2013.[185] McCarthy was not only the third-ranking Republican in the House, but also from a district that was 37 percent Hispanic and reliant on immigrant labor. The congressman was also known for his ties to Silicon Valley and known to be in favor of expanding the H-1B program. To make their case, pro-immigration groups besieged McCarthy with rallies, phone calls, letters, and other targeted efforts in late 2013. In response, however, the opposition mobilized to nullify the pro-immigration campaign. The local Tea Party staged its own events and met with the congressman to emphasize their opposition. NumbersUSA also got involved, asking its activists to phone and message McCarthy's offices with their concerns. As the president of NumbersUSA, Roy Beck, later explained, "We did react on McCarthy, because he was truly in play."[186] In the end, individuals close to the congressman told the *Wall Street Journal* that the intensely conflicting pressures made him wary of taking a lead on the issue.[187]

The larger test would come in 2014. In early January, the *New York Times* reported that Speaker of the House John Boehner "was said to back change in immigration," particularly after he hired Rebecca Tallent, an immigration adviser who had worked for Senator John McCain.[188] On January 30, Boehner and other Republican leaders released a set of immigration "principles" that not only included calls for stronger border security and employment verification, but also for granting legal status (if not citizenship) to some undocumented immigrants.[189] Boehner then organized a discussion of the principles at a House Republican retreat.

The opponents were ready. The fact that Boehner would be raising the issue of immigration at the retreat was known in advance, so opponents were lobbying before it began. NumbersUSA encouraged House Republicans who were "cold on Boehner" and considering skipping the event to attend and to put pressure on the speaker.[190] The Tea Party Patriots organized a robocall campaign, which resulted in 41,046 protest calls from voters to ninety members of Congress, according to the organization.[191]

Other conservative groups, including Heritage Action for America and ForAmerica, also exhorted their supporters to call Congress. ForAmerica claimed that it left more than 5,500 voicemail messages saying that voting for the Senate-passed immigration reform bill was "turncoat" and would have consequences.[192] In the end, Boehner was unable to get his principles adopted, and the pressure tactics of opposition groups were an important reason why. As one higher education lobbyist who was closely involved recalled, "There were a lot of members of Congress who wanted to get to yes on [immigration reform], including within the House Republican caucus, but based on the building pressure and the building pressure and the building pressure, couldn't get there."[193] Or as Scott Corley of Compete America later put it, many House Republicans "adhered to the principle that it was not their job to lead their constituents to the truth but to acquiesce in populism."[194] A week after the retreat, Boehner declared that it was unlikely the House could pass immigration reform in 2014.[195]

Although immigration reform appeared dead for the year, there was a revival of interest in the spring. Analysts had earlier noted that Boehner might wait until May or June to push the issue, since most House Republicans would have completed their primary elections by that point.[196] And in March, Boehner reportedly told donors and industry groups that he was "hell bent" on passing immigration reform in 2014.[197] Boehner also publicly taunted House Republicans for resisting reform, eliciting rebukes from various opponents.[198] Yet if there was still a chance of passing a bill that year, it evaporated in June. On June 10, the House majority leader, Eric Cantor, lost his seat in a primary challenge from an obscure Tea Party figure, David Brat, who had attacked Cantor for supporting Boehner's push for immigration reform.[199] Cantor's loss stunned the GOP and put the final nail in the coffin for a comprehensive bill that year.

With Congressional action stalled, President Obama was determined to make some reforms to the system unilaterally. Toward this end, White House officials consulted closely with advocates for both the HTC and undocumented immigrants in the second half of 2014 regarding potential executive actions. High-tech advocates generated dozens of ideas as to how the executive branch could make the system friendlier from their perspective.[200] The administration was generally sympathetic, and it was willing to be "very creative," as one White House official later put it.[201] Yet it was also constrained by the relevant legislation, which made some changes—such as changes to the number or allocation of H-1B visas—infeasible.

Starting in November 2014, the Obama administration undertook a series of executive actions that offered something to both legalization advocates and the HTC.[202] For the former, the administration announced a program to defer deportation for some parents of U.S. citizens or legal permanent residents. It also expanded a program to defer deportation for some undocumented immigrants who had come to the United States as children. On the high-tech front, the administration permitted certain spouses of H-1B visa holders to work, making life easier for those families, provided the H-1B holder had applied for a green card. The administration also offered portable work authorization to high-skilled workers with approved green card applications, making it easier for H-1B workers to be promoted or change jobs. The administration increased the extension period for STEM graduates in the OPT program to twenty-four months, as noted earlier, and used the government's "parole authority" to make it easier for foreign startup entrepreneurs to work in the United States. Other measures remained out of reach, however. The administration thought carefully about allowing the "recapture" of unused EB visas from previous years, a move strongly favored by high-tech employers like Microsoft. Administration officials ultimately determined that the move was not clearly available under the law, however.[203]

Indeed, the order and its aftermath underlined the limited power of the executive branch to alter U.S. immigration policy. The deferred deportation measures were immediately challenged by Texas and twenty-five other states, and their implementation was subsequently halted as the case was considered in the judicial system. The HTC, meanwhile, remained deeply frustrated with the U.S. immigration system. The H-1B cap and the system of visa allocation remained unchanged, and the green card backlog remained unresolved. "Everyone has been saying nice things for years and has nothing to show for it," said Compete America's Scott Corley after Obama's plans were announced in late 2014. "We want to see actions. We want to see concrete outcomes. And Congress has to do the heaviest lifting."[204] In short, the battle was far from over.

CONCLUSION

The fluctuating U.S. policy toward skilled immigration over the past two decades reflects a shifting tug-of-war between the HTC and its opponents. High-tech firms and universities generally support liberal policies and have fought for them repeatedly. The success or failure of these efforts, in turn, has depended upon the nature of the resistance faced by these interests from other groups. When the HTC has encountered resistance primarily from labor, it has largely prevailed, although it had to accept temporary measures in 1998 and 2000 and the persistence of a cap more generally. In contrast, when the HTC has encountered resistance from large and well-organized citizen groups, it has failed to increase the level of openness, regardless of labor's stance.

Alternative explanations for U.S. policy offer unsatisfying accounts. According to the strategic hegemon hypothesis outlined in chapter 2, we should expect the United States to restrict collaboration with China and India if doing so were feasible and less costly for it than for them. It is unclear how this approach would explain the rising and falling H-1B cap since the mid-1990s, however. The feasibility of adjusting the cap has not clearly changed over time. Although the cost of restricting collaboration has changed, it has not done so in a way that could explain changes in U.S. policy. From the early 2000s to 2008, the cost of limiting the H-1B cap increased for the United States as the U.S. economy recovered from the bursting of the dot-com bubble. Yet the H-1B cap fell substantially during this period, particularly after the temporary increases passed in 2000 expired. More generally, if some sort of realpolitik strategy were driving U.S. policy, one would expect the preferences of the executive branch to play an important role in driving the fluctuations of the H-1B cap. Yet this has not been the case. The Clinton administration was unenthusiastic about raising the H-1B cap in 1998, yet it was raised later that year nonetheless. The Bush and Obama administrations sought to raise the H-1B cap as part of broader immigration reforms, but both were unsuccessful. In short, it is hard to see the rising and falling H-1B cap as a reflection of national strategy.

Public opinion toward immigration also struggles to explain the fluctuations in U.S. policy. In September 1998, a Harris poll of 1,000 U.S. adults indicated that 82 percent were opposed to raising the H-1B cap substantially, and only 16 percent were in favor.[205] Despite the results of this poll, which were immediately publicized by the IEEE-USA, the first increase in the H-1B cap passed the House of Representatives shortly thereafter. Public opinion also fails to offer a compelling explanation for the reduced U.S. openness to skilled immigration since 2004. Indeed, public opinion toward skilled immigration softened between the late 1990s and 2008. One survey conducted in late 2007 and early 2008 found that only 40 percent of respondents were opposed to increases in skilled immigration.[206] In 2013, a Gallup poll found that 76 percent of respondents were in favor of expanding visas for short-term skilled workers.[207] The same poll also found large majorities in favor of creating a pathway to citizenship for some undocumented immigrants if certain conditions were met. In short, public opinion cannot explain why the United States expanded skilled immigration between 1998 and 2004 but has not done so since.

What about the rising and falling fortunes of individual political parties? The preferred party hypothesis predicts that the power of labor and citizen groups to resist the HTC will vary widely over time, depending on the relative strength of the major political parties. This view offers some degree of insight into U.S. policy. Labor clearly had more ability to contend with the HTC in 2007 (when the Democrats controlled both Houses) than it did in 2006 (when the Republicans controlled both). Anti-immigration citizen groups, in contrast, have been most formidable when trying to influence Republican-controlled chambers. Overall, however, labor has typically struggled to prevent the HTC from accomplishing its goals, even when the Democrats have controlled the White House. In contrast, citizen groups have repeatedly stymied the HTC since 2005.

The status quo hypothesis predicts that the HTC will be most successful when defending extant policy and least successful when trying to change it. This approach, too, struggles to explain the varying outcomes discussed in this chapter. The HTC was attempting to change U.S. immigration laws in both time periods examined here, but it was far more successful in the first period than in the second. The status quo bias offers no insight into these divergent outcomes.

The two coalition strength hypotheses also perform poorly in this case. The first of these predicts that the HTC will be most successful when both its corporate and academic wings are actively engaged. The importance of such collaboration was evident in the 2000 and 2004 victories, when research universities actively supported ICT firms' push for an increase in H-1B visas. Even so, higher education remained engaged after 2004, but efforts to liberalize skilled immigration during this period failed. The second coalition strength hypothesis predicts that the HTC will be most successful when its corporate wing allies with other business interests. This was most clearly the case in the battles over comprehensive immigration reform between 2006 and 2014, when high-tech firms allied with other business groups (as well as Latino groups and other organizations). Yet none of these efforts was successful. The key reason, of course, was that the opposition during this period was far more potent. It is variation in the opposition faced by the HTC, rather than variation in the strength of its coalition, that explains the outcomes observed in this chapter.

4

THE OPEN DOOR

Foreign Students

T HE GLOBALIZATION OF higher education is thriving. Between 1975 and 2012, the number of students abroad at the post-secondary level soared from 0.8 million to 4.5 million.[1] The trend has been especially impressive in the twenty-first century: Annual growth in international student mobility averaged 7 percent between 2000 and 2012.[2] The United States, in turn, has embraced this trend. For decades, the U.S. government has placed no cap on the number of foreign students who can study at U.S. universities, and the United States has enrolled more foreign students than any other country in the world. By 2015/16, more than 1 million international students were enrolled at U.S. colleges and universities.[3] Since the late 1990s, China and India have emerged as the most important sources of these students. China became the top sender in 1999/2000, eclipsing Japan, only to be surpassed by India in 2001/02.[4] China regained the lead once again in 2009/10.[5] Overall, foreign students remain a relatively small share of U.S. undergraduate enrollment, but they have taken on a substantial share of the graduate population, particularly in S&E fields.[6]

In this chapter, I explore the policies and politics behind U.S. openness to foreign students in the early twenty-first century. In the first section of the chapter, I describe how the United States has been open to foreign students, particularly since the mid-1960s. The result has been dramatic increases in the number of Chinese and Indian students in the United States over the past few decades. In the second section, I explain how the HTC, led by higher education groups, has succeeded in sustaining such openness, particularly in the absence of organized opposition. The analysis here first considers U.S. policymaking in the immediate aftermath of the terrorist attacks of September 11, 2001, when members of Congress raised the possibility of placing a temporary ban on foreign students. The analysis then considers U.S. policymaking in the years following the attacks, when the admission of foreign students and scholars became a considerably more difficult process to navigate. In both cases, higher education groups accomplished their goals, not merely because of energetic lobbying efforts but also because of the absence of organized opposition.

AN OPEN DOOR

The United States has explicitly admitted foreign students to study at American educational institutions since the Immigration Act of 1924. In the post–World War II era, students have been admitted through one of three visa programs. The F-1 visa, the most heavily used type, is designed for foreign students pursuing a full-time academic education. The M-1 visa is designed for vocational students. The J-1 visa is designed for visitors participating in a cultural exchange program designated by the Department of State. Some of these international visitors are students, but some are professors, postdoctoral researchers, medical graduates, and au pairs, among others. Unlike the H-1B visa, these visa categories have no annual cap.[7]

In the absence of a cap, recent decades have seen unprecedented numbers of foreign students enter the United States. In 2014, the Department of State granted 644,233 F-1 visas, 332,540 J-1 visas, and 11,058 M-1 visas.[8] Of these three categories, China and India loom largest in F-1 visas. In 2015, China accounted for 12 percent of J-1 visas, 18 percent of M-1 visas, and a whopping 43 percent of F-1 visas. India accounted for 2 percent of J-1 visas,

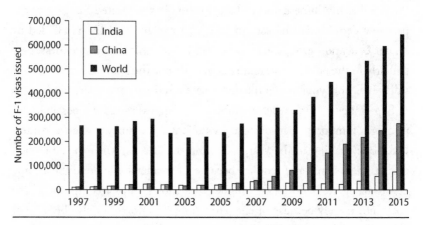

FIGURE 4.1 Number of F-1 visas issued by the U.S. Department of State, 1997 to 2015. *Source*: U.S. Department of State, "Nonimmigrant Visa Issuances by Visa Class and by Nationality," *Travel.State.Gov*, accessed September 9, 2016, https://travel.state.gov/content/visas/en /law-and-policy/statistics/non-immigrant-visas.html.

6 percent of M-1 visas, and 12 percent of F-1 visas. As shown in figure 4.1, the F-1 shares for China and India were similar until 2007, when each country accounted for between 10 and 15 percent of the total. China's soaring share since then largely reflects the rising number of Chinese students who have enrolled in undergraduate programs in the United States. Chinese students have gone in greatest numbers to prominent public universities like the University of California, Los Angeles, the University of Texas at Austin, the University of Washington, and the "Big Ten" schools.[9]

Foreign students in the United States are particularly prominent in S&E fields at the graduate level. Despite the influx of undergraduates from overseas, and China in particular, foreign students made up only 4.2 percent of total enrollment in U.S. higher education in 2013/14, although on some campuses foreign students have made up more than 15 percent of incoming classes in recent years.[10] If we focus on graduate S&E education, foreign students represent a more impressive share of the total student population.

Foreign students on temporary visas received 31 percent of S&E doctorates granted by U.S. universities in 2000, and by 2013, this figure had climbed to 37 percent.[11] In some fields, particularly engineering, computer science, and economics, such students received half or more of all doctorates awarded. China and India, meanwhile, have become by far the largest sources of foreign graduate students in S&E fields. In fact, of the 209,020 foreign graduate students enrolled in S&E fields at U.S. universities in November 2014, China accounted for 33 percent of the total, and India accounted for 35 percent.[12]

The specific disciplines favored by Chinese and Indian graduate S&E students are fairly similar. As of 2014, engineering was the most popular discipline for students from both countries, while computer science was second. Mathematics and the physical sciences were the third and fourth most popular specialties, respectively, for Chinese students. Biology and the physical sciences were the third and fourth most popular fields, respectively, among Indian students. The specialty in which Chinese and Indian students were most prominent was computer science, where they comprised a stunning 83 percent of all foreign graduate students.[13]

As they have become more numerous, foreign graduate students have become significant contributors to academic innovation in the United States. A 2008 study estimated that a 10 percent increase in the number of foreign graduate students at U.S. universities would increase the number of patent applications by 4.7 percent, university patent grants by 5.3 percent, and non-university patent grants by 6.7 percent.[14] A subsequent study reported that both foreign and American graduate students make significant and positive contributions to academic innovation and argued that restrictions on the entry of high-quality students, regardless of nationality, would have harmful effects on university innovation.[15] In addition, foreign graduate students often stay in the United States following graduation, and this is particularly true of Chinese and Indian students. As recently as 2013, more than 80 percent of Chinese and Indian doctorate recipients on temporary visas intended to stay in the United States after graduation.[16]

THE FIGHT FOR OPENNESS AFTER SEPTEMBER 11

The sustained openness of the United States to foreign students is not simply the result of government on autopilot. Instead, it reflects the active efforts of the HTC to maintain an open system, as well as a lack of organized opposition. Indeed, the HTC (particularly the academic wing) has developed a strong interest in the flow of students from abroad. For many U.S. universities, foreign students have become an important source of revenue. In fact, U.S. education exports grew from $13 billion in 1999 to nearly $36 billion in 2015.[17] This revenue stream became particularly important after 2008, when the sagging American economy led to funding cuts for many public universities. Foreign students in S&E disciplines are also an important source of well-educated labor at many U.S. research universities, as described earlier. Technology companies also have a keen interest in the flow of S&E students from abroad. Foreign students may work in the United States after graduation for a limited period of time through the Optional Practical Training Program, as noted in chapter 3, after which they may be sponsored for a work visa such as the H-1B. Some companies even cultivate relationships with promising foreign S&E students while they are still in school in hopes of hiring those students after graduation.[18]

Although both the corporate and academic wings of the HTC have an interest in international students, the academic wing has normally led the lobbying on this issue. Whereas foreign students are a primary concern for universities, they are a secondary concern for technology companies. As a result, as one higher-education representative put it, business tends to "free-ride" on the lobbying of universities with regard to foreign students, which allows business lobbyists to focus their resources on other issues.[19] For their part, university interests have traditionally been represented in Washington by a range of groups, as noted in chapter 2. Particularly prominent are the "Big Six": the American Association of Community Colleges, the American Association of State Colleges and Universities, the American Council on Education (ACE), the Association of American

Universities, the Association of Public and Land-Grant Universities, and the National Association of Independent Colleges and Universities. ACE functions as the umbrella organization and has coordinated meetings of the Washington Higher Education Secretariat, which includes roughly fifty higher education–related groups as members. The National Association for Foreign Student Affairs (NAFSA), meanwhile, represents international education professionals. The Council on Governmental Relations (COGR) represents and advises research universities, medical centers, and research institutes conducting at least $15 million annually in federally sponsored research. Major research universities also typically have Washington offices and conduct their own lobbying operations.

FENDING OFF THE BAN

Although U.S. universities have developed a strong interest in flows of international students, the Big Six were relatively quiescent on this subject in the 1990s. As two scholars stated in 1998, American higher education had yet to become "an articulate, well-organized advocate before the U.S. government on behalf of initiatives to promote international programs and academic exchange."[20] Indeed, when funding for the Fulbright Program was cut by 20 percent in 1996, there was no great outcry from American academia.[21] When Congress passed legislation that same year to establish a monitoring system for international students, it was NAFSA that led the opposition to the new system.[22] Larger higher education organizations, in contrast, either played a supporting role or were not involved.[23]

The terrorist attacks of September 11, 2001, had dramatic implications for the politics surrounding flows of foreign students to the United States— and also for the activity of higher education organizations. Eighteen of the nineteen Al-Qaeda hijackers entered the United States on tourist and business visas.[24] One of them, Hani Hanjour, entered the United States on an F-1 student visa to study English but never attended class and was never reported missing. In addition, two other hijackers successfully applied to have their visas converted to vocational student status.[25] These revelations led to a pronounced shift in favor of tighter immigration policies following the attacks, and student visas were a particular point of concern. As the *Washington Post* reported in October 2001, "The recent terrorist attacks have radically altered the immigration debate, replacing an agenda of amnesty with proposals to remilitarize U.S. borders, severely limit student visas and increase tracking of foreigners on American soil."[26] Or as the vice-president of ACE, Terry Hartle, later put it, "Everything was up for grabs."[27]

Nonetheless, the HTC would be largely successful over the next several years in its bid to sustain international student flows, particularly since it faced little in the way of organized opposition. The first challenge it faced was particularly dire: the prospect of a temporary ban on foreign students altogether. On September 27, 2001, Senator Dianne Feinstein proposed a six-month moratorium on student visas to give the Immigration and Naturalization Service time to fix the system's many problems.[28] Feinstein's proposal was not the most severe: Longer moratoriums were proposed in the House of Representatives. But her proposal was particularly important, since the senator was the chair of the Senate Judiciary Subcommittee on Technology, Terrorism, and Government Information. Feinstein also called for funding for an electronic student-tracking system, new admission procedures, greater security at ports of entry, and other measures. The proposed moratorium was taken "very seriously" by higher education officials, who feared that such measures could be implemented in the new political climate.[29]

The result was a round of "intense lobbying," as the *Boston Globe* put it, by higher education officials.[30] On October 2, the president of ACE, David Ward, wrote a letter to the Senate Judiciary Committee, cosigned by twenty-nine other higher education organizations, in response to Feinstein's proposal.[31] The letter "vigorously oppose[d]" the moratorium. It argued that the ban was "not in the nation's strategic or economic interest" and that "the loss to the United States in terms of intellectual accomplishment, goodwill, and lost economic activity [would] be enormous." It also argued that the ban would be ineffective, since student visas were only 2 percent of all visas granted overall. The letter further argued that it would take "decades" to undo the damage that even a short-term ban would impose. The letter's points were driven home in meetings between Feinstein and higher education officials, including representatives from public and private universities in the senator's home state of California.[32]

In response, Feinstein's proposal was quickly shelved. On October 5, shortly after meeting with university leaders, the senator said that the moratorium "may not be necessary."[33] Her press secretary said that the senator had decided to "pull back" following her dialogue with higher education representatives.[34] Feinstein did so with the understanding that universities would support the introduction of a monitoring system for foreign students. As her office explained, "If we can get cooperation from schools with regard to student visa reporting requirements, the moratorium will not be necessary."[35] Yet ACE had never opposed the idea of a student monitoring system.[36] Rather, the council had expressed concerns about how the development of such a system would be funded. Feinstein was now proposing full federal funding for this purpose—a move for which ACE and other associations expressed "strong support."[37] The retraction of the proposed moratorium was described in the press as "a major victory" for higher education.[38]

The victory was not surprising. Simply put, the universities faced little organized opposition as they pressed their case, and what little opposition they did face was too weak and too slow to matter. The clearest opposition came from the anti-immigration group, the Federation for American Immigration Reform (FAIR). As noted in chapter 3, FAIR was relatively small and poorly placed to mobilize grass-roots resistance. On October 24—more than two weeks after Feinstein had backed down— the president of FAIR, Daniel Stein, appeared on CNN's *Crossfire* and argued against what was now a fait accompli.[39] But FAIR was not only too late. It was also quite lonely in its opposition, and no grass-roots campaign to pressure Feinstein emerged.[40]

Indeed, FAIR had little support from other groups. Domestic student organizations, notably, were more concerned about student freedoms than foreign competition for enrollment slots after September 11. In an interview in late October, for example, the United States Student Association's legislative director expressed concern about the "web of suspicion cast over international students" and charged that student privacy was "clearly being compromised."[41] FAIR also had little support from other anti-immigration groups. Mark Krikorian of the Center for Immigration Studies, an anti-immigration think tank, summed up the situation in mid-2002: "Actually, we haven't done much on foreign students, and . . . actually a big hole, I think, in our examination of the whole of immigration policy is that almost nobody has looked at foreign students."[42]

The HTC faced no serious opposition as it fought the idea of a moratorium on student visas. Years later, in fact, ACE executives did not recall any opposition from other organized groups.[43]

TOWARD A WORKABLE SYSTEM

There were more challenges to come in the wake of September 11. Although there was no moratorium, the executive branch was now far more cautious in the way it approached foreign students, and the issuing of visas for students and academic visitors became a considerably more vexed process. In response, the HTC engaged with both the executive branch and Congress to address its concerns. Whereas fighting the moratorium had entailed putting pressure on Congress, the HTC now found itself engaged in an ongoing dialogue with the executive branch over how the administration should attempt to ensure national security. As noted in chapter 2, this kind of engagement lies beyond the scope of the theory I am testing in this study. Even so, I explore it here to gain some insight into the HTC's relative effectiveness under such conditions, particularly when unopposed by other groups.

To begin, the executive branch's caution after September 11 was evident on a number of fronts. The denial rate for F-1 visas rose from 20 percent in fiscal year 2000 to more than 27 percent in fiscal year 2002, and it remained above 25 percent in 2003.[44] The denial rates for Chinese and Indian F-1 and J-1 applications were especially high, reaching 42 percent for China and 43 percent for India in 2003.[45] Some Chinese students who were denied visas went so far as to stage protests outside the U.S. embassy in Beijing.[46] In general, applications were more often delayed, sometimes causing students to start their studies late or to miss their programs entirely, and returning students often encountered challenges in trying to re-enter the country. Academic institutions found it difficult to predict which applications would require more time, making the delays difficult to manage.[47] More rigid enforcement of existing rules also created new challenges for foreign students after they arrived in the United States.[48] These challenges, combined with the weaker U.S. economy and greater competition for students from other countries, undermined the United States' position as the leading destination for overseas study.[49] From 2001 to 2004, in fact, applications for F-1 visas dropped by nearly 100,000. Applications for graduate study declined by 28 percent in 2004 and then a further 5 percent in 2005, with notable declines in applications from Chinese and Indian students.[50] Graduate engineering programs were especially hard hit, given that these were heavily reliant on inflows of Chinese and Indian students.[51]

These challenges were of considerable concern both to universities and to technology companies. In late 2003, the president of Yale University, Richard Levin, stated that eliminating barriers to student visas had become the top priority in Washington for his and other American universities.[52] For their part, technology companies worried that a reduced number of foreign S&E students in the United States would limit their access to foreign talent. "We're concerned this could hurt the pipeline of masters and Ph.D.s in engineering in the years to come," one manager at a high-tech multinational told the *Far Eastern Economic Review* in November 2002. "This could have a serious impact on industry."[53] Or, as the president of the Information Technology Association of America, Harris Miller, told CNN earlier in the year, "Half of all graduate students in the math and science programs [in the United States] are foreign students. When a company is looking for the best and brightest, particularly people with advanced degrees, master degrees and Ph.D.s, frequently, many of those candidates are born abroad."[54]

Despite the shared concern, it was clearly higher education's role to take the lead in tackling this challenge, given not only the priority universities gave to attracting foreign students, but also their direct involvement in the admissions process. Higher education groups took action on multiple fronts. On May 12, 2004, the heads of twenty-four higher education, science, and engineering organizations jointly published their "Statement and Recommendations on Visa Problems Harming America's Scientific, Economic, and Security Interests." The statement charged that "visa-related problems are discouraging and preventing the best and brightest international students, scholars, and scientists from studying and working in the United States." It predicted that "the misperception that the United States does not welcome international students, scholars, and scientists will grow" in the absence of action to correct it. It warned that the damage to the country's "higher education and scientific enterprises, economy, and national security would be irreparable." The statement proceeded to single out six specific problems and made recommendations regarding how each could be addressed. To drive the points home, university presidents highlighted the statement in testimony before the Senate Committee on Foreign Relations in October, and the statement was republished in full in the official record.[55]

Higher education groups also engaged in a variety of lower-profile efforts with Congress. ACE organized a trip to London for congressional staffers so they could better understand the process through which visas were granted. London was chosen not only for reasons of proximity, but also because more student visas were granted there at the time than in any other location, including those of many Indian students.[56] NAFSA also made an effort to enlist Congress. When a university was having particular trouble with delayed applications, NAFSA would inform members of Congress from the university's state.[57] Higher education groups also engaged in direct dialogue with the Department of State and the executive branch more generally about student visas. As the vice-president of ACE, Terry Hartle, later recalled, "We were pushing on them" to solve these problems.[58] NAFSA also worked with the Department of State, as well as other departments involved in the interagency review process, to identify problem areas and expedite the process. NAFSA would also call attention to particular cases in which an application was taking a particularly long time. In some cases, this was helpful because officials in Washington were not always aware of delays at the consular level.[59]

While these efforts were underway, higher education groups also high-lighted problems with another aspect of the foreign student program: the new monitoring system. In late 2002, the Department of Homeland Security established the Student and Exchange Visitor Information System (SEVIS). Schools were required to start using it for new students and visitors as of February 15, 2003, and for all students and visitors begin-ning August 1, 2003. In March 2003, the president of ACE, David Ward, complained before Congress that the executive branch was not being sufficiently responsive to higher education's concerns with regard to the new system.[60] Ward proceeded to outline a range of problems with SEVIS. Due to its rapid rollout, the system suffered from a variety of technical and administrative problems, ranging from data losses and bizarre irregu-larities to inadequate efforts to certify educational institutions.[61] NAFSA subsequently worked closely with the Department of Homeland Security to improve the system.[62]

As higher education officials worked to address these various challenges, the organized opposition they faced was negligible. FAIR did propose an annual ceiling on student visas on its website during this period, arguing that foreign students created greater competition for domestic students in admissions and funding.[63] There was no grass-roots campaign in pursuit of this position, however. While NumbersUSA might have helped FAIR in this regard, it was still quite small in the early 2000s, and the group took a different view of foreign students in any case. The president of NumbersUSA, Roy Beck, would later say his organization did not seek to limit the number of foreign students studying in the United States, calling higher education "one of our great exports."[64] To be sure, there were some individual critics of the foreign student program. In 2002, for example, the Harvard economist George Borjas questioned the value of admitting so many foreign students in an article published in the *National Review*.[65] The article criticized the foreign student program on national security grounds and charged that it was corrupting university admis-sion standards, among other concerns. Borjas also authored a report for the Center for Immigration Studies that was highly critical of the foreign student program.[66] Yet these isolated arguments failed to arouse more general opposition. In fact, when higher education officials testified before Congress about foreign students in March 2003 and October 2004, the panels featured no critics of the program.[67]

Facing so little opposition, higher education would succeed in achieving its goals. From 27 percent in 2002, the refusal rate for F-1 visas fell to 20 percent in 2005 and 2006.[68] The Department of State also made a concerted effort to reduce visa application delays. It turned out that the time required to adjudicate a visa request depended heavily on whether the applicant had to undergo an interagency security check known as Visas Mantis. First introduced in 1998, the process was designed to ensure that visas were denied to individuals considered likely to promote the transfer of controlled technologies to countries deemed state sponsors of terrorism or other "countries of interest." After September 11, the number of such checks skyrocketed, and the review process became much slower as a result. In 2003, the General Accounting Office (GAO) noted that the Mantis reviews took an average of sixty-seven days to complete.[69] Reviews for applicants from China, India, and Russia were particularly slow. By late 2004, however, the average Mantis review took just fifteen days, thanks to a concerted effort to streamline the process.[70] Moreover, the denial rate remained low: only 2 percent of all cases.[71] There was also a broader effort to expedite student visas more generally: In July 2004, the Department of State issued a cable indicating that applicants for F, J, and M visas were to be given priority in scheduling, since such applicants often faced deadlines for arrival in the United States.[72] The problems with SEVIS, too, were ameliorated. By October 2004, in fact, the president of NAFSA, Marlene Johnson, described SEVIS as a secondary concern since "the remaining issues are largely technical."[73] By 2005, the GAO described the system as "improved."[74] Although work continued on SEVIS thereafter, it became a back-burner issue.

With the situation stabilizing in 2005, the HTC was suddenly confronted with a different challenge to the flow of foreign students and researchers: export controls. Traditionally, academic research had been exempted from the government's licensing requirements if the work qualified as "fundamental research." In March 2005, the Department of Commerce sought public comments on a proposed rule that would have greatly increased the licensing burden on universities.[75] The rule aimed in particular to broaden "deemed export" regulations, which govern the transfer of technology to foreign nationals in the United States.[76] It redefined the use of controlled equipment to include simply operating it, whereas in the past, multiple criteria had to be satisfied for use to be considered to have occurred. The rule also would have applied controls based on a foreign national's country of birth, rather than his or her most recent country of citizenship or residence. This latter change reflected particular concern that Chinese individuals who had become citizens or permanent residents of U.S. allies—and thus subject to less stringent controls—could be exploited by Chinese intelligence services.[77]

Higher education groups reacted swiftly. In late June, the Association of American Universities (AAU) and COGR submitted nineteen- and sixteen-page letters, respectively, to the Department of Commerce that raised a number of issues and concerns.[78] These included the impact on the image of U.S. universities as welcoming to foreign scholars and students; the importance of foreign students and scholars in U.S. scientific research; the extent of the licensing burden the proposed rule would create; the unlikelihood of improving national security through these measures; and the negative impact on scientific research itself. ACE and other higher education groups submitted critical reactions as well. There was also a wider public reaction. A total of 307 public comments were made on the proposed rule, more than 200 of which were submitted by academic scientists and researchers, and these were decidedly negative as well.[79]

In May 2006, the Bureau of Industry and Security within the Department of Commerce announced that it was withdrawing the proposed rule.[80] The definition of use would remain unchanged, and deemed export controls would not be applied based on country of birth. This was a clear victory for higher education. There was one setback, however. The Bureau stipulated that a license could be required for the use of controlled equipment in the conduct of research even if the reporting of research results was not subject to control. This went against the positions taken by the AAU and COGR, which saw the conduct and reporting of research as inseparable. Even so, the practical impact of this change was very limited, given the withdrawal of the other proposals. Indeed, higher education officials were quite pleased with the outcome overall. As an internal COGR memo put it, "For the most part the announcement represents good news for the university research community."[81] The licensing burden would remain manageable, and the image of U.S. higher education would not be tarnished by new and burdensome technology controls. When a license was required, moreover, the Department of Commerce was extremely unlikely to deny it.[82]

The vice-president of ACE, Terry Hartle, later recalled that in 2006, the policy environment surrounding foreign student admissions became "stable" once again.[83] Against this backdrop, foreign student admissions would reach new heights over the next several years. Whereas applications for F-1 visas fell to 311,497 in 2004, they bounced back impressively thereafter, as shown in Figure 4.1. In 2008, the State Department received 504,607 F-1 visa applications, of which it granted 340,711.[84] Applications to attend U.S. graduate schools rebounded as well, growing by 12 percent in 2005/06 and 9 percent the year after.[85] The rising number of applications actually overwhelmed the State Department in 2008, leading to substantial processing delays and renewed complaints from higher education groups, but once again the concerns were addressed.[86] Although applications slowed in 2009, as the American economy struggled with the global financial crisis, growth resumed once again in 2010. In 2013, the State Department issued more than half a million F-1 visas for the first time ever.

CONCLUSION

In the years following the September 11 attacks, the HTC—led by higher education groups—worked to ensure that the United States remained open to foreign students. In the immediate aftermath of the attacks, higher education officials confronted a proposed moratorium on student visas that would have halted inflows of foreign students in the short term and raised serious questions about the U.S. commitment to international education over the long term. In reacting to this proposal, the HTC faced negligible organized opposition from other groups. Its efforts were thus highly successful. Over the next few years, the influx of foreign students would nonetheless decline, in part a result of higher rejection rates for visas, processing delays, and the negative perceptions that these problems generated abroad. In 2005, the possibility of new export controls emerged. To address these challenges, higher education groups engaged directly with the executive branch, which was forced to balance national security concerns with the realities of globalizing education and academic research. Such engagement between the HTC and the executive branch lies outside the scope of the theory outlined in chapter 2. Nonetheless, the fact that the HTC was so successful in this effort is striking, and it demonstrates that high-tech interests have the potential to shape the executive's approach to national security quite powerfully.

What about alternative explanations for U.S. policy toward foreign students? The strategic hegemon hypothesis would suggest that the United States has acted strategically. More specifically, this view would explain U.S. policy by noting how much it would cost the United States to exclude foreign students and by noting that the United States does not possess a monopoly on the provision of higher education. Although these broad points seem reasonable, they overlook important facts. First, the United States has many options available to it that would be far less costly than a blanket ban or cap on foreign students. U.S. policymakers could limit the number of students from some countries but not from others, for example, ensuring that the costs of such restrictions fell disproportionately on the student-sending state. Second, although the United States does not possess a monopoly on higher education, it does predominate in the upper echelons of science and engineering education. For 2014/15, the *Times Higher Education* rankings listed fourteen U.S. universities in the top twenty worldwide for engineering and technology and thirteen U.S. universities in the top twenty worldwide for the physical sciences.[87] For graduate study in particular, there are opportunities in the United States that do not exist elsewhere. Third, a strategic approach to foreign students would imply efforts to create more employment opportunities for such students in the United States following graduation. Yet as chapter 3 made clear, such opportunities have diminished since 2004. Taking these points together, it is difficult to explain U.S. policy toward foreign students in strategic terms.

The public opinion hypothesis also struggles to illuminate the outcome in this case. Following the September 11 attacks, U.S. public opinion took a decided turn against immigration. A poll in October 2001, for example, found that 58 percent of Americans wanted immigration levels decreased—a striking jump from 41 percent just four months earlier.[88] In this context, a temporary moratorium or an annual cap on foreign students would hardly have been unpopular. Yet the HTC was highly successful against this backdrop of unfavorable public sentiment.

The preferred party hypothesis is not relevant in this case. There was no significant opposition to the HTC from labor or citizen groups, so we cannot evaluate how the relative power of different parties affected their ability to resist the HTC.

The coalition strength hypotheses are particularly unhelpful with regard to U.S. policy toward foreign students. The first of these predicts that the HTC will be most successful when both of its wings are actively engaged. In this case, the higher education wing took the lead, while the corporate wing played a supporting role in the background. Further, broader business interests did not take an active role on the issue. But even though one wing was much more active than the other in this case, the HTC was strikingly successful. The second coalition strength hypothesis—that the HTC will be most successful when its corporate wing allies with other business interests—is simply not relevant here.

The most useful alternative explanation would appear to come from the status quo hypothesis. The United States has not traditionally limited the number of foreign students at U.S. universities, so enacting even a temporary ban on such students would have represented a dramatic departure from extant policy. Even so, the HTC did not succeed here because of the difficulty of enacting a ban on foreign students. It succeeded because it dissuaded a key legislator from even attempting to enact such a ban. In addition, as the processing of student visas became a serious challenge following the September 11 attacks, the task facing the HTC evolved from defending a favorable status quo to trying to change what had become an unfavorable status quo. The HTC succeeded in this task as well, even though it required engaging with the executive branch on a matter of national security.

If higher education groups took the lead on student visas following September 11, 2001, the corporate wing of the HTC would take charge on a different front in the early 2000s: global R&D. I consider that story in the next chapter.

5

THE (MOSTLY) OPEN DOOR

Global R&D

T HE FLOW OF brainpower across national borders is a crucial ingredient in the globalization of innovation. Yet the way in which firms and universities perform R&D is also globalizing. Multinationals' R&D centers have sprouted up in increasingly disparate locations across the globe, and there is growing cross-border collaboration among firms, universities, and individual researchers. In the past, this activity primarily involved developed countries. Today, however, developing countries—particularly China and India—play increasingly prominent roles in global R&D. These new roles owe much to the United States. U.S. firms are easily the most prominent R&D investors in both China and India. The United States is also the main partner for China and India in R&D alliances. U.S. openness, then, has played a crucial role in integrating China and India into global R&D.

Global R&D is a challenging phenomenon for the United States to regulate. By its very nature, high-tech R&D is complex and requires a high degree of expertise to understand. The ambitions of U.S. scientists and engineers, and the openness of American society more generally, also work against constraints on cross-border communication and collaboration with foreign colleagues. Partly for these reasons, the United States has been and remains largely open to global R&D. Nonetheless, there is an important aspect of global R&D that the U.S. government regulates in a variety of ways: foreign investment. The Committee on Foreign Investment in the United States (CFIUS) reviews incoming investments, particularly those that could result in control of a U.S. business by a foreign person, and can recommend to the president that a transaction be blocked on national security grounds.[1] U.S. investments overseas, in turn, are subject to a variety of U.S. laws and regulations, particularly those regarding taxation, trade, and export controls. Although global R&D is challenging to regulate, then, it is hardly unregulated, particularly in the realm of foreign investment.

For two reasons, this chapter focuses on U.S. regulation of outward foreign direct investment (OFDI). First, this is the policy domain in which the most serious challenge to global R&D has been mounted to date. Since the early 2000s, in fact, there have been a variety of efforts to change U.S. laws relevant to OFDI in order to limit the movement of R&D and other service jobs overseas. In contrast, although interest in reforming CFIUS has grown in recent years, no major effort to restrict incoming investments has been made. Second, and even more important, it is in the realm of OFDI that high-tech interests have been engaged. Challenges to OFDI worry leading high-tech firms in particular, given their extensive international investments and operations. They have therefore responded to these challenges energetically. In contrast, the review of incoming investments through CFIUS has not traditionally been a concern for U.S. high-tech firms as a group, although it is sometimes a concern for individual companies.[2] For that reason, the politics surrounding incoming investments are beyond the purview of this study.

I begin the chapter with an overview of key U.S. policies relevant to OFDI. I focus on several broad policies toward foreign investment, since it is within this wider context that global R&D has been considered and debated. I explain how the United States has traditionally been quite open in this regard but also that a range of concerns has arisen in recent years, particularly regarding the impact of offshoring on U.S. employment and technological leadership. Despite these concerns, U.S. lawmakers have taken very few of the opportunities they have had to alter U.S. policies. To explain this outcome, I then delve into the politics behind U.S. policy to show how the corporate wing of the HTC has worked with other business allies to fend off anti-offshoring legislation at both the federal and state levels. I also explore the opposition faced by high-tech firms in this endeavor, particularly from organized labor. Because labor has been essentially alone in this fight, high-tech has usually (if not always) prevailed.

THE (MOSTLY) OPEN DOOR

The United States has a longstanding tradition of openness to OFDI, and this openness rests on several distinct pillars. The first is tax policy. In the first half of the twentieth century, tax credit and tax deferral became important parts of U.S. policy toward OFDI.[3] Tax credit allows corporate taxes paid to foreign governments to be counted against U.S. corporate tax liabilities. Tax deferral allows tax liabilities on foreign income to be deferred until the income is repatriated. The rationale for deferral is that it allows U.S. companies to compete in low-tax foreign countries at parity with local competitors. Critics have charged, however, that U.S. companies use deferral to lower their effective tax rates and that it encourages them to expand abroad rather than at home.[4] These tax provisions are highly pertinent to the global operations of leading U.S. technology companies. Companies including Apple, Dell, Hewlett-Packard, IBM, and Microsoft have been called "prime examples" of firms that enjoy high foreign profits and low effective tax rates.[5]

The second pillar is trade policy. With the emergence of global innovation and production networks in recent decades, another key aspect of U.S. OFDI policy has been its openness to goods and services produced abroad. U.S. openness in ICT in particular is embodied in support for the Information Technology Agreement under the World Trade Organization (WTO), including the expansion of this agreement in recent years.[6] Trade openness is important to OFDI because U.S. firms sometimes rely on foreign subsidiaries, and international networks more generally, to produce for the U.S. market, particularly in ICT. For this reason, labor unions have focused not only on international tax policy, but also on trade-related measures as a means of combating offshoring.[7]

The third pillar is U.S. export controls, particularly those in the dual-use realm. The Commerce Control List consists of ten categories of dual-use technologies, including electronics, computers, and telecommunications and information security, with different countries subject to varying levels of control. Even so, controls in key areas of ICT, including high-performance computers and semiconductors, have been progressively liberalized since the 1990s.[8] In other areas, such as software development and cloud computing, few controls exist. The system is also increasingly focused on the end use of controlled technologies, so that civilian use may be permitted whereas military use may be restricted. With a few exceptions, therefore, current U.S. controls are not a major constraint on the globalization of R&D in ICT.[9] As one knowledgeable IBM executive observed with regard to China in particular, export controls remain "relevant" in some cases but generally are "not a tier-one factor" for his firm in deciding whether to do in-country R&D.[10] More important considerations include the availability of local talent and the security of intellectual property. Other companies, such as Google, have avoided creating controls on new technologies by choosing not to develop products for the U.S. military, focusing on civilian applications instead.[11]

In short, U.S. policy toward international taxation, trade, and export controls is broadly supportive of OFDI in general and the development of foreign R&D in particular. This openness, in turn, has allowed U.S. multinationals to expand their R&D activity abroad over the past few decades. In 2013, U.S. companies spent more than $49 billion in R&D overseas through their foreign affiliates, as noted in chapter 1. Whereas most (61 percent) of this spending was directed to Europe, the share directed to China and India is growing. Indeed, the list of U.S. ICT firms investing in R&D in these two countries now reads like a who's-who of high-tech corporate America: Apple, Cisco, Dell, Google, Hewlett-Packard, IBM, Intel, Microsoft, Oracle, and Qualcomm, among others. In fact, U.S. companies have emerged as the primary foreign investors in R&D in both China and India. In China, the United States and Japan are the two leaders, but the gap is large: U.S. companies spent 2.7 times as much as Japanese companies did on R&D in China between 2005 and 2013.[12] U.S. companies also lead in India by a wide margin. Between 2003 and 2009, the latest years for which data are available, U.S. firms accounted for 53 percent of foreign spending on R&D in India, whereas Germany was a distant second at 8 percent.[13]

Although concerns have existed in the United States about OFDI since the 1970s, worry about the movement of service jobs in particular, including those in R&D, emerged in the early 2000s. In November 2002, a report by Forrester Research predicted that 3.3 million services jobs in the United States would move offshore by 2015. The report added that the ICT industry would "lead the initial overseas exodus."[14] This report helped generate a sense of alarm about the future of U.S. technological competitiveness, particularly with regard to China and India. As Intel's chief executive put it, "The structure of the world has changed . . . the United States no longer has a lock on high-tech, white-collar jobs."[15] Some worried about the dangers of offshoring, while others decried "offshore outsourcing." As other commentators pointed out, these were distinct phenomena: *Offshoring* refers to companies moving operations to their foreign subsidiaries, whereas *offshore outsourcing* refers to contracting out business functions to foreign companies in overseas locations.[16] Whatever the phraseology used, the sense of alarm was real. In March 2004, a Gallup poll reported that 61 percent of Americans worried that they, a relative, or a friend might lose a job because an employer was moving the job overseas. The same poll reported that 58 percent of Americans said that this issue would be "very important" when deciding how to vote for president later that year.[17]

Some worried about what offshoring meant for how the United States was positioning itself vis-à-vis China and India. A report released by Senator Joseph Lieberman's office in May 2004, for example, argued that the United States was "losing out" to these two countries in high-tech sectors:

The offshoring of facilities, labor, capital, technology, and information not only hurts our workers, but also threatens the backbone of our knowledge-based economy. Emerging nations such as China and India have realized that technological leadership leads to economic prosperity. Their governments are committed to attracting business investments, technology transfer, and knowledge inflow into their countries through industrial policies, subsidies, and business incentives.[18]

The report further stated that "the sheer size of China['s] and India's popu-
lations and their far lower costs of living mean that their low wages will
put pressure on the U.S. workforce for a very long time to come."[19] Since
then, concerns about what offshoring means for the competitive position of
the United States have persisted. As a prominent commentator observed in
June 2016, "While moving operations offshore initially means a loss of jobs,
over time it leads to the movement of more assets, including research and
development capabilities, out of the country, which, over the long term is
crippling to our competitiveness."[20]

In response to these concerns, the potential benefits of offshoring have
been trumpeted. In 2004, the chair of the Council of Economic Advisers,
Gregory Mankiw, called offshoring "the latest manifestation of the gains
from trade that economists have talked about" for centuries and thus "a good
thing."[21] Although Mankiw's remarks proved highly controversial, even
economists who highlighted the potential for disruption, such as Alan
Blinder, disparaged the idea of erecting barriers to offshoring.[22] Some
argued that the United States stood to gain from trade in services and
that it should seek to liberalize such trade—rather than trying to stop it.[23]
Others argued that the U.S. multinationals more active overseas tend to be
more active in the United States as well, whether one focuses on employment, sales, capital expenditures, or R&D.[24] Even some critics of offshoring,
such as the Institute of Electrical and Electronics Engineers-USA (IEEE-
USA), noted that there was a danger of trying to do too much to restrict it.
"Offshoring is a complex issue," said the president of the IEEE-USA, John
Steadman, in March 2004. "It's clear there are some areas where if you pro-
hibit any level of offshoring, companies could fold."[25]

In the face of this complex and ongoing debate, U.S. policy has remained
largely open to offshoring. This has not been for lack of ideas about how
to restrict it: There have been a variety of proposals at both the federal and
state levels to limit offshoring since the early 2000s. Few of these have
become law, however. In January 2004, for example, Congress passed legislation limiting the offshoring of federal contract work in that fiscal year, but
a proposal to make the provision permanent was later defeated. Since 2010,
Congress has also considered several bills to limit offshoring more generally,

but none has become law, with one limited exception. Many state governments have also considered measures to limit offshoring, and most (if not all) of these have been defeated as well.

Let us now turn to the politics behind the persistent U.S. openness to offshoring.

FIGHTS OVER OFFSHORING

Following the growing controversy in late 2002, the U.S. government began to face the offshoring phenomenon in 2003. As it did so, lawmakers would confront lobbying from two opposing sides. On the pro-openness side, technology companies took the initiative early on, since much of the controversy over offshoring concerned the ICT industry. The Information Technology Association of America (ITAA) was particularly active in trying to shape the debate. The group engaged the media with the message that the impact of offshoring on U.S. employment had been exaggerated and that it was vital for U.S. companies to remain competitive and profitable. The group also stressed the danger that other countries would retaliate if the United States became less open. As the president of the ITAA, Harris Miller, explained on National Public Radio in March 2004,

> Many of our companies are generating a huge percentage of their profits by selling abroad. When we're winning the game, when we're driving down the field, when we're being successful, it would be absolutely insane for us to say, "Let's stop the game and change the rules and disadvantage ourselves." That would be the height of folly.[26]

The ITAA joined forces with other business groups as it pressed its case, and in early 2004, these groups created a new organization: the Coalition for Economic Growth and American Jobs. The group included roughly 200 business associations, including the ITAA, the American Bankers Association, the Business Roundtable, the National Association of Manufacturers, and the U.S. Chamber of Commerce.[27] The prestige, wealth, and capacity for organization of the group was obvious, and its power was noted in the media at the time.[28]

On the other side of the debate was organized labor, including both the IEEE-USA and the AFL-CIO. Labor leaders understood the magnitude of the challenge they were facing as they worked to limit offshoring. Thea Lee, chief international economist at the AFL-CIO, described the opposition: "This is a very, very wealthy group of businesses, and if they choose to spend a lot of resources [on this issue], that will be something to be reckoned with."[29] Lee hoped that support from public opinion and the fact that companies were "on the wrong side of the issue" would allow labor to win the day, or at least some of the battles. Still, it was a lopsided contest. As one writer put it in 2004, "The forces opposed to anti-offshoring legislation are better organized, richer, and more sophisticated in the ways of leveraging influence. And they have made blocking restrictions a top priority."[30]

The political struggle over offshoring resembles the struggle over the H-1B visa program in the late 1990s. Technology companies lobbied in favor of openness, whereas organized labor worked against it, with the balance of power favoring tech. In this case, in fact, the balance of power was even more heavily tilted against labor, as technology companies cooperated more closely with other businesses that shared their concerns. The disparity in power would play out in lopsided legislative fights at both the federal and state levels. Although organized labor has managed to win a few battles, the HTC and its allies have won the war.

THE FIGHT AT THE FEDERAL LEVEL

Starting in 2003, the AFL-CIO began working with Congress on measures to combat offshoring.[31] While the labor giant naturally worked most closely with Democratic lawmakers, there was concern on both sides of the aisle. These efforts, combined with wider public concern, began to pay off toward the end of the year. Congress naturally focused on the one aspect of business that it could influence most easily: government procurement.

In October, the Senate voted 95 to 1 to approve an anti-offshoring amendment to an appropriations bill for several departments and agencies. The amendment required federal contract work awarded during fiscal year 2004 under Office of Management and Budget (OMB) circular A-76 to be done in the United States.[32] The bill was passed and signed by President Bush in January 2004.

This particular measure did not arouse a great deal of opposition from the HTC. The bill's impact was likely to be limited, and companies were concerned that fighting it would tarnish their image as good corporate citizens.[33] There was concern, however, that the bill would lead to broader restrictions. The president of the ITAA, Harris Miller, explained the concern following the passage of the procurement restrictions in January: "Companies . . . are concerned as even if the impact of this bill is limited, opponents of offshoring are going to push for greater protection."[34] Business leaders did not believe that the government would find a way to stop offshoring, but they did worry that government could make it considerably more difficult and expensive.[35]

The fear that greater restrictions would be proposed was well founded. In February 2004, Senator Christopher Dodd of Connecticut introduced the United States Workers Protection Act, following a meeting between Dodd's staff and AFL-CIO specialists.[36] Whereas the restrictions imposed by the earlier amendment had been limited to 2004, Dodd's bill made them permanent. Dodd's bill also extended the restrictions to state contracts with federal funding.[37] The bill was clearly responding to concerns about China and India in particular: Many European countries would have been unaffected by the bill because they were among the twenty-seven signatories of the WTO agreement on procurement.[38] Dodd subsequently agreed to adjust the bill's language in negotiations with other senators from both parties, including Max Baucus, John McCain, and Mitch McConnell (who was then majority whip). Although the new language created some exceptions, the core idea of a permanent ban on offshoring federal contract work remained.[39]

Dodd's effort received enthusiastic support from a coalition calling itself the Jobs and Trade Network (JTN). Although the coalition described itself as an alliance of business and labor groups working against off-shoring, it was essentially a labor initiative. The fifteen sponsors included representatives from various parts of the AFL-CIO; the Paper, Allied-Industrial, Chemical and Energy Workers (PACE) International Union; and the United Steelworkers.[40] The few business groups supporting the JTN, in contrast, were small players like the Pennsylvania Manufacturers' Association and the Manufacturing Alliance of Connecticut. Some of these had chosen to support the JTN out of frustration with their inability to influence national business groups, such as the National Association of Manufacturers and the U.S. Chamber of Commerce, which were unsympathetic to their concerns.[41] While the IEEE-USA was not part of the JTN, it too supported the anti-offshoring effort. In mid-March, the group released a position paper arguing that U.S. government procurement rules should favor work done domestically and should "restrict the offshoring of work in any instance where there is not a clear long-term economic benefit to the nation or where the work supports technologies that are critical to our national economic or military security."[42]

Technology companies, as part of the Coalition for Economic Growth and American Jobs, made a bigger effort to defeat Dodd's proposal than they had made against the more limited bill in January. Even so, they did not wish for their opposition to tarnish their image or alienate the U.S. government as a customer. The ITAA and its allies were therefore careful in how they went about defeating the amendment. Working with allies in Congress and the administration, the business coalition quietly informed key lawmakers how their businesses would be affected if Dodd's proposal were to become law.[43] The president of the ITAA, Harris Miller, also asked Indian IT companies to remain quiet and let their U.S. counterparts take charge of the lobbying.[44] The Coalition also worked to reframe the public debate. Meeting with officials from the White House, the Department of Commerce, and the Office of the U.S. Trade Representative, Coalition leaders suggested using the term *worldwide sourcing* rather than *offshoring*, and they stressed the importance of recalling other moments in history when fears emerged that jobs would disappear owing to new technology or foreign competition only to prove unfounded.[45] The ITAA itself sponsored a study entitled "The Impact of Offshore IT Software and Services Outsourcing on the U.S. Economy and the IT Industry," which painted offshoring in a more positive light. The report stated that offshoring was not the main cause of job losses in the IT services sector and that the number of IT jobs in the United States would increase even with offshoring.[46] Coalition members also warned about the consequences of restricting commerce with other countries. Officials at AeA (formerly the American Electronics Association) argued that foreign companies could react to protectionist U.S. policies by reducing their investments in the United States.[47]

The Coalition's efforts were not enough to prevent passage in the Senate. In May, the Senate passed a tax bill that included Dodd's proposal as an amendment by a vote of 92 to 5.[48] The Coalition would ultimately succeed in making the amendment die a quiet death, however. In June, the House approved its own version of the tax bill by a vote of 251 to 78, but without language similar to that of the Dodd amendment.[49] House and Senate negotiators met to reconcile the two bills in the fall, and industry would move to exert its influence here. As the *Washington Post* reported, "Under heavy lobbying pressure, GOP negotiators generally sided with the House version."[50] Paul Almeida, the president of the AFL-CIO's Department for Professional Employees, later recalled that industry made "a tremendous push to get [the Dodd amendment] dropped."[51] The effort was successful: The Dodd amendment did not appear in the final bill that emerged from the conference.

Although efforts to restrict offshoring at the federal level receded after 2004, the issue made a comeback following the election of Barack Obama to the White House in 2008. This time, the focus was not on government procurement but on tax incentives. This was not a new idea. In March 2004, the AFL-CIO executive council argued that "federal tax policies that encourage shipping U.S. jobs overseas must be replaced by tax incentives focused on job creation here."[52] Senator John Kerry had also repeatedly raised the issue during the 2004 presidential campaign, singling out "Benedict Arnold companies" that were engaging in offshoring.[53] On the campaign trail in 2008, Obama would raise the issue of overseas taxation and offshoring once again.[54]

Upon taking office, it was unclear what Obama would do. In February 2009, Obama promised to end "the tax breaks for corporations that ship our jobs overseas" before a joint session of Congress.[55] The following month, however, Obama tried to lower expectations. In a town hall meeting at the White House, Obama acknowledged that offshoring was not going away. "Not all of these jobs are going to come back," the president stated. "What we've got to do is create new jobs that can't be outsourced."[56] In May, however, the administration launched an initiative entitled "Leveling the Playing Field: Curbing Tax Havens and Removing Tax Incentives for Shifting Jobs Overseas."[57] Among other things, the administration sought to require companies to pay taxes on their overseas income before taking deductions related to overseas expenses.[58] The administration projected that the changes, which would be implemented in 2011, would raise federal revenue by $210 billion over ten years.

To fight the proposal, a number of business groups formed the Promote America's Competitive Edge (PACE) coalition. The coalition included not only the National Association of Manufacturers and the U.S. Chamber of Commerce but also high-tech groups, including the Information Technology Industry Council, TechAmerica, TechNet, and the Technology CEO Council.[59] TechAmerica had roughly 1,200 members; the group was a merger of the American Electronics Association, the Cyber Security Industry Alliance, the Government Electronics and Information Technology Association, and the ITAA. Technology firms were particularly alarmed because of their extensive investments overseas.[60]

Following a brief but intense lobbying effort, Obama shelved the plan in the fall of 2009.[61] In private meetings with administration officials, the business community emphasized that the plan would do little to advance Obama's top priority: stabilizing the economy in the midst of the global financial crisis. Moreover, they argued, the plan would be very difficult to implement, it would affect many businesses in unwelcome ways, and it would do little to reduce offshoring.[62] The resistance from technology firms, many of which had supported Obama in 2008, was particularly noteworthy. Given the administration's emphasis on the economy, there was particular reluctance to tinker with a sector with such great potential for growth.

As one official later recalled, "We were not going to expend a lot of political capital on something they were going to fight us on, especially if it didn't look like it would actually have a positive impact on the economy."[63]

The issue did not simply disappear, however. In February 2010, the Obama administration's budget request proposed changes to taxes on overseas corporate income that would raise $122 billion over ten years—a considerable retreat from the previous year's target of $210 billion but still a very significant sum.[64] Although these changes would not be adopted, several more focused attempts to reform overseas corporate taxation from abroad were made. In late May, for example, Senate Finance Chair Max Baucus and House Ways and Means Chair Sander Levin jointly introduced the American Jobs and Closing Tax Loopholes Act. The bill included a variety of tax and spending measures, including provisions to raise taxes on venture capitalists and increase revenue from overseas corporate income by $14 billion over ten years.[65] Both measures were anathema to Silicon Valley, which rallied against it. Yet if Silicon Valley hated the bill, labor loved it. Whereas the IEEE-USA kept a low profile, the president of the AFL-CIO, Richard Trumka, lauded the legislation as "cracking down on tax loopholes for millionaire hedge fund managers and on corporations that ship our jobs overseas."[66] Labor also had public opinion on its side. "It's incredibly dangerous to be on the wrong side of this issue," said the Democratic pollster Geoff Garin. "There's a belief that the most important thing we can do to prevent unemployment is to keep jobs from going overseas."[67] With labor and public sentiment in favor, the Democratic-controlled House passed the bill 215 to 204 on May 28, with voting largely along party lines.

The bill would founder in the Senate, however. In mid-June, the PACE coalition sent a letter to all members of the Senate explaining that the foreign tax provisions would compromise U.S. companies' competitiveness abroad. Republican senators Kit Bond and John Thune took up the cause, with Bond offering an amendment to remove the foreign tax provisions from the bill.[68] Although the amendment was defeated, the Democrats now confronted a Republican filibuster. The Senate's Democratic leadership attempted to end debate on the bill three times in June, but each attempt failed—the last by only three votes. Following the last defeat, the majority leader, Harry Reid, tabled the bill.[69]

Labor and its Democratic allies would finally record a small victory toward the end of the summer. In early August, several Democratic senators amended a bill originally introduced to modernize the Federal Aviation Administration (FAA). The revised bill provided $26 billion in funding to states in order to prevent teacher layoffs and support Medicaid payments.[70] The amendment included several measures to pay for these outlays, including changes to overseas taxation that would generate roughly $10 billion over ten years. The bill enjoyed strong support from labor, particularly since teachers' unions were now intensely interested in it. At the same time, cracks were beginning to show in the HTC's coalition. On August 2, twenty-two companies—including the technology firms Hewlett-Packard, National Semiconductor, Qualcomm, and Texas Instruments—signed a letter to Senate leaders endorsing changes to taxation of overseas income if the revenue were used to fund the lapsed R&D tax credit. Although the FAA bill did not reinstate the credit, the letter served to reveal disagreement within the business community over the importance of maintaining the status quo on overseas tax policy. As the *Wall Street Journal* reported, the letter marked "the first indication of a splintering within businesses and could embolden Democrats."[71] The Democrats pounced on the opportunity. On August 4, the Senate agreed to end debate on the bill by a vote of 61 to 39, with the moderate Republicans Olympia Snowe and Susan Collins parting company with the rest of their party and voting in favor.[72] The Senate passed the bill by the same margin on August 5. The House signed off on August 10, and the president signed the bill the same day.[73] The victory was real, but small: Having once aimed at $210 billion in new revenue, the anti-offshoring forces were now settling for $10 billion.

The Democratic leadership would attempt to build on this small success in the fall. Indeed, in the run-up to the November 2010 elections, it became apparent that voters remained highly anxious about jobs moving abroad. One poll asked respondents to identify the most important action government could take to speed economic recovery. The top result was a tie: reducing outsourcing to foreign countries and increasing skills.[74]

Democratic lawmakers moved quickly to take advantage of this development. In September 2010, the Creating American Jobs and Ending Offshoring Act was introduced in the Senate.[75] The bill provided tax incentives for companies to hire U.S. workers to replace foreign employees. It also prohibited firms from receiving deductions taken in connection with the closing of an operation in the United States if the company was starting or expanding a similar business overseas. Last, the bill repealed overseas tax deferral for companies that reduced or closed a trade or business in the United States and started or expanded a similar business overseas for the purpose of exporting products to the United States. Once again, the bill was supported by labor, particularly the AFL-CIO, but resisted by business groups. The latter included TechAmerica and the U.S. Chamber of Commerce. The president of TechAmerica, Phil Bond, wrote a letter to the entire Senate membership arguing that "increased tax costs . . . would ultimately be borne by U.S. employees, customers, or shareholders."[76] Given the amount of resistance the bill was expected to encounter, some Democratic supporters privately acknowledged that it would be difficult to enact.[77] In September, the bill died after the Senate voted 53 to 45 not to cut off debate on the measure.

The AFL-CIO continued to fight against offshoring after the vote, hoping to influence the November elections. The president of the AFL-CIO, Richard Trumka, accused Republicans in the Senate of taking "one last slap at working families before adjourning" and stressed that "working people are facing a choice" in November.[78] In early October, the AFL-CIO launched a new website that allowed voters to identify companies in their zip codes that had offshored jobs, based in part on whether workers had received federal trade adjustment assistance for losing jobs to foreign competition.[79] With its affiliated group, Working America, the AFL-CIO also published a report entitled "Sending Jobs Overseas: The Cost to America's Economy and Working Families," which stressed the extent of the phenomenon. Among other measures, the report called for renewed efforts to pass the Creating American Jobs and Ending Offshoring Act.[80]

The outcome of the midterm elections, however, was a terrific setback for organized labor, dashing its hopes of getting an anti-offshoring bill through Congress. The Republicans gained a stunning sixty-three seats in the House, taking back control, as well as six seats in the Senate, reducing the Democrats' majority. The Republicans also made major gains in state legislatures, increasing their ability to influence the redrawing of Congressional districts. As a result, critics of offshoring lowered their sights regarding what could be accomplished through federal legislation.[81]

To be sure, there have been some anti-offshoring efforts since 2010. These have typically been election-year efforts focused on "grandstanding bills," as one close observer has described them, and they have not been seen as likely to pass.[82] In 2012, for example, the Democratic senator Debbie Stabenow of Michigan introduced the Bring Jobs Home Act, which labor groups supported. The bill offered companies tax incentives to move jobs to the United States from overseas, and it denied a tax deduction for outsourcing expenses incurred in relocating a U.S. business to a foreign location. The bill stalled in the Senate in July, when a motion to cut off debate received only fifty-six votes.[83] The bill was reintroduced in 2014, but it suffered the same fate.[84]

Despite the concerns about offshoring that emerged in the early 2000s, anti-offshoring legislation has generally struggled to become law at the federal level, although labor has had a few victories. Overall, the HTC and its allies have succeeded in fending off legislation that would limit offshoring. This was true even in 2009/10, when Democrats controlled the White House and both houses of Congress. Yet the fight between labor and the HTC was not limited to the federal level; it would also play out in the states.

THE FIGHT IN THE STATES

As the HTC was fighting the Dodd amendment in 2004, it was also fending off a wave of anti-offshoring legislation at the state level. In 2003/04, according to one tally, more than 200 anti-offshoring bills were introduced in more than forty state legislatures.[85] In 2005/06, 190 such bills were introduced. Organized labor worked nationwide to move this legislation forward. "The AFL-CIO has mobilized support for such bills around the country," the *Washington Post* reported, "urging not only outright bans on overseas contracting but also an end to tax incentives that encourage overseas work and measures that force contractors to disclose where their employees are located."[86] As it did at the federal level, labor led the fight against offshoring in the states.

The HTC and its allies in the business community fought back vigorously. In July 2004, the president of the ITAA, Harris Miller, told the press that "offshoring is our most time-consuming issue by far . . . we're actively fighting all of these bills."[87] Or, as Miller put it on National Public Radio, "We're like the volunteer fire department in the middle of a raging drought. I mean, there [are] embers blowing up all over the place, and we just have to keep running around and tamping them down."[88]

In the end, the ITAA and its allies were largely successful. Although hundreds of anti-offshoring bills were introduced, only a handful became law. In 2003/04, for example, only Alabama, Colorado, Indiana, North Carolina, and Tennessee passed laws that gave preferences in the awarding of government procurement work to contractors in the United States.[89] Moreover, the bills that passed were typically quite restrained. The Indiana law gave price preferences to in-state firms ranging from 1 to 5 percent. The North Carolina law established a preference for domestic products and services, but only if this entailed no sacrifice in price or quality. Alabama passed a resolution that merely encouraged state and local entities to use in-state services, with no mandate or requirement on procurement decisions. Labor leaders were disappointed at the outcome. "It's incredible the stranglehold the corporate interests have on this issue," said Marcus Courtney, the president of the Washington Alliance of Technology Workers.[90]

What explains this lopsided outcome? Simply put, the Coalition for Economic Growth and American Jobs was not only wealthy and well organized, but also savvy in the ways of shaping and blocking legislation. Working at the state level, the group employed several tactics in particular: enlisting large employers to tell lawmakers that they would be hurt by proposed restrictions, warning that taxpayer costs would increase if offshoring were limited, and playing for time to run out the clock on legislative sessions. The Coalition also stressed how legislation could run afoul of trade agreements and reduce foreign investment. These "blocking and tackling" techniques, as one U.S. Chamber of Commerce executive called them, were conveyed to Coalition members through business conferences and other means.[91]

The ITAA leadership later recalled several keys to success at the state level.[92] First, because many companies already had a lobbying effort in virtually every state, the most efficient approach was to convince corporate members that stopping anti-offshoring legislation ought to be one of their top legislative priorities. A large number of companies telling lawmakers that the issue was important to them would send a strong message. Second, it was important to prevent the offshoring debate from becoming a highly public shouting match played out through the media. The more intense the public debate, the harder it would be to convince lawmakers to let anti-offshoring bills die quietly. Third, some companies were savvy enough to make adjustments that would help to defuse the issue. Companies with call centers abroad, for example, designed their systems so that customers who sounded uncomfortable could immediately be switched to a call center in the United States.

All in all, then, the technology companies and their allies succeeded in the states not only because of their terrific wealth and impressive organization, but also through a nationwide campaign that leveraged these resources intelligently. When the HTC was unsuccessful, moreover, it was sometimes the result of unusual circumstances. In 2005, for example, New Jersey passed what was arguably the country's strictest anti-offshoring law, one that prohibited government contract work from being performed outside the United States.[93] Because the law marked an unusual setback for the HTC, the circumstances surrounding its passage deserve additional scrutiny.

The bill was first introduced in the New Jersey State Senate by Senator Shirley Turner in 2002. In December, the bill passed the State Senate unanimously, 40 to 0. This development got the attention of the ITAA, which then mounted what the *Wall Street Journal* called a "fierce lobbying campaign" to stall the bill in the State Assembly.[94] In addition, India's National Association of Software and Service Companies, a New Delhi trade group representing 850 companies, hired the lobbying firm Hill & Knowlton to fight the New Jersey bill (as well as other anti-offshoring bills). These efforts seemed to pay off. The bill stalled in the State Assembly in 2003 and was not given a hearing or vote.[95] Union members swamped the Assembly with thousands of angry emails calling for action, but to no avail.[96]

In 2004, Senator Turner reintroduced the bill, hoping that the passage of federal anti-offshoring legislation in January would boost its prospects.[97] As the president of the ITAA, Harris Miller, later said, the federal bill gave "a certain amount of aid and comfort to people at the state level who are trying to promote similar legislation: 'The feds did it, why can't we?'"[98] In June, Turner's bill succeeded in passing the Senate once again, although the 29-to-5 vote in favor was not the unanimous endorsement it had received in 2002. Once again, the bill stalled in the Assembly. In response, Turner appealed to Governor James McGreevey, a Democrat, for help. McGreevey was a logical choice: The president of the AFL-CIO, John Sweeney, had described McGreevey's administration as one of the most pro-labor state governments in the country.[99] Turner found the governor uninterested, however. "He said, 'What can I do?'" she later recalled. "He more or less shrugged his shoulders."[100] Turner later speculated that McGreevey was concerned about opposition to his reelection effort from the business community.[101] Once again, the bill appeared to be stuck in limbo.

This would change in August. That month, McGreevey confessed to having an extramarital affair with another man and announced his intention to resign on November 15. The deferred resignation removed the need for a special election, allowing State Senate President Richard Codey, a fellow Democrat, to succeed McGreevey as governor. McGreevey also indicated that he wished to undertake some final actions that would cement his legacy.[102] In September, he would take one of those, signing an executive order that restricted the offshoring of government contract work. The order allowed state agencies to hire firms employing overseas workers only if another firm using U.S. workers could not perform the service, the cost of using U.S. workers would create economic adversity for the state, or the state treasurer declared the contract would best serve the public interest.[103] Although not as strict as Turner's bill, the measure marked a significant departure from the governor's earlier stance.

If McGreevey hoped to have the last word on the matter, he would be disappointed, however. "With all due respect, we need a law not an executive order," Turner said after the order was issued.[104] "Governors come and governors go, and new ones can wipe out the executive orders of their predecessors with the stroke of a pen." On November 15, McGreevey stepped down, and Richard Codey became acting governor. The transition, as Turner later put it, was "very helpful" in the effort to pass the anti-offshoring law.[105] Codey represented a relatively liberal district, and he had already supported the bill in the State Senate.[106] On March 14, 2005, the Assembly passed the bill, 68 to 5. The bill was signed into law by Codey on May 5.

In sum, New Jersey's strict anti-offshoring law emerged in quite unusual circumstances. Governor McGreevey had once been touted as a rising star within the Democratic Party.[107] The scandal that forced his resignation, however, turned him into a lame-duck governor who was no longer concerned with reelection or national office—and thus much more difficult for the HTC to influence. Only in the wake of the scandal did McGreevey issue his executive order to restrict offshoring. His resignation shortly thereafter, in turn, facilitated the passage of the anti-offshoring law.

After 2006, interest in anti-offshoring legislation receded at the state level. The number of bills introduced in state legislatures on offshoring or outsourcing (as it is often called) peaked in 2006 at 209 but fell below 100 per year after 2007, as shown in figure 5.1. This is a rough indicator, of course, based on whether a bill synopsis included the word *offshoring* or *outsourcing*. Not all of these bills were anti-offshoring measures, and some anti-offshoring bills did not contain the word *offshoring* or *outsourcing*. Moreover, there were still some noteworthy efforts to restrict offshoring at the state level after 2006. In 2010, for example, Tennessee passed a law forbidding the state from contracting out services that would entail the use of overseas call centers.[108] That same year, the governor of Ohio, Ted Strickland, issued an executive order prohibiting the use of public funds for services provided offshore.[109] Even so, the apparent decline in state-level efforts to restrict offshoring since 2006 is striking, and the trend is congruent with data derived from media coverage.[110] The decline suggests that organized labor concentrated its anti-offshoring efforts following the 2008 elections at the federal level.

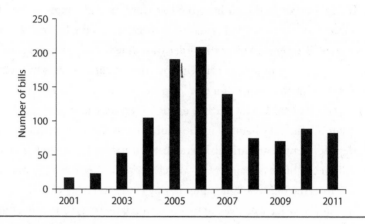

FIGURE 5.1 Number of state bills with the word *offshoring* or *outsourcing* in the bill synopsis, 2001 to 2011. *Source*: LexisNexis State Net database, accessed August 4, 2015.

CONCLUSION

Since the early 2000s, the primary challenge to global R&D in the United States has been anti-offshoring legislation. Particularly from 2003 to 2006, but also from 2009 to 2010 at the federal level, organized labor led a series of attempts to reduce the extent of offshoring, if not stop it altogether. In general, these efforts focused on tax and trade policy, rather than export controls. Although the focus was not only on high-skill jobs, many of the proposals would have complicated the expansion of overseas R&D.

High-tech firms have been worried by the anti-offshoring movement, and they have responded with highly effective efforts to limit its impact. Indeed, working with other business groups, high-tech firms have been largely successful in their fights with labor. High-tech firms suffered a small setback at the federal level in early 2004, when Congress passed legislation that limited the offshoring of government contract work for the remainder of that year, but they succeeded in preventing this change from becoming permanent. High-tech firms also suffered a small set-back in August 2010, when disarray in their ranks helped produce a small change in the international corporate tax regime. Technology companies subsequently helped defeat federal anti-offshoring bills in 2010, 2012, and 2014, however. In the states, many anti-offshoring bills were introduced between 2003 and 2006, but the HTC and its business allies prevented most of them from becoming law. The most restrictive law that was passed, in New Jersey, came about in unusual circumstances that made it difficult for technology firms to succeed. As a result, the United States has remained mostly open to offshoring—an outcome that fits well with the theory outlined in chapter 2.

How do alternative explanations fare in this case? The strategic hegemon hypothesis predicts that the United States would restrict collaboration with China and India if doing so were feasible and less costly for it than for them. In terms of feasibility, restricting collaboration is certainly more difficult in this case than is the case in immigration. Whereas legal immigration can be monitored and regulated, corporate foreign investment is more difficult to influence, particularly when large corporations keep ample funds overseas. Nonetheless, the U.S. federal and state governments do have tools at their disposal to address offshoring, but they have generally not used them. Infeasibility, then, is far from the whole explanation. What about costs? While experts will disagree about the economic impact of the various bills discussed in this chapter, it is clear that the Obama administration viewed several of these bills in positive terms, and some of them promised to increase federal revenue. In contrast, anti-offshoring legislation would have had no clear upside for China or India. It is thus difficult to see the distribution of costs as the deciding factor here.

The public opinion hypothesis is clearly lacking in support in this case. Although public opinion has typically been on the side of labor, it has been of limited political utility. As noted earlier, a sizable majority of the public was worried about offshoring in 2004, but it was in this year that the Dodd amendment was defeated. Public opinion was also favorable to anti-offshoring legislation in the run-up to the 2010 midterm elections, but efforts to pass anti-offshoring legislation at this time were defeated as well.

The preferred party hypothesis also struggles to find much support. In this view, labor's ability to resist the HTC should be considerably greater when a progressive party (the Democrats) predominates than when a generally pro-business party (the Republicans) does. Yet labor was largely unsuccessful throughout this time period, regardless of which party was more prominent. This was true even in 2009/10, when the Democrats controlled the executive branch and both houses of Congress. Although the Democrats did attempt to pass the Creating American Jobs and Ending Offshoring Act in 2010, the effort was defeated in the Senate—with four Democrats and one Democrat-turned-independent (Joseph Lieberman) voting against cloture.

The status quo hypothesis is more useful. In fact, it can be seen as a complement to the interest-group analysis I have undertaken in this chapter. High-tech firms were largely successful not only because they outmuscled organized labor, but also because they were defending the status quo while labor was trying to change it. The failure of federal anti-offshoring bills to pass in 2010, 2012, and 2014 is a good example of this dynamic. These bills all foundered in the Senate, where the supermajority needed to move forward made it particularly easy for high-tech interests to block change. Even so, there were also instances—such as the shelving of Obama's tax reform initiative in 2009—in which high-tech firms managed to prevent legislation from being introduced in the first place. This suggests that the status quo bias is part of the explanation in this case, but not the entire explanation.

The evidence I have presented in this chapter has very different implications for the two coalition strength hypotheses. The first maintains that the HTC will be most successful when both its corporate and academic wings are actively engaged. This is poorly supported by the evidence here: Despite the academic wing being uninvolved, the HTC was largely successful. In contrast, the second hypothesis—concerning the degree of support from other businesses—is much better supported. High-tech firms clearly allied with other business groups, including the National Association of Manufacturers and the U.S. Chamber of Commerce, and this support helps explain why the HTC was quite successful in this case. Nonetheless, it is worth noting that even this impressive business coalition suffered some defeats. This outcome stands in contrast to that observed in the battle over foreign students described in the preceding chapter. In that case, the HTC had fewer allies, but it also faced less opposition, and its victory was more decisive.

CONCLUSION

T HREE BROAD FINDINGS emerge from the preceding chapters in this book. The first, quite simply, is that innovation is globalizing. This is particularly evident in the massive cross-border flows of brainpower that have now become commonplace, whether one considers skilled workers or students in S&E disciplines. R&D activity is also globalizing. This is apparent in growing cross-border collaboration, whether one considers R&D alliances, venture capital activity, university partnerships, or scientific co-authorship. This is also evident in the growing corporate investment in overseas R&D centers, although U.S. multinationals still do most of their R&D at home. These changes have allowed developing countries, particularly China and India, to play roles in innovation they have never played before. These changes have also compelled the United States to consider how it should collaborate with emerging powers in an economic arena it has long dominated.

The second finding is that U.S. openness to global innovation is driven by the very organizations that embody the country's technological leadership. Together, U.S. high-tech firms and research universities have important shared interests in openness to flows of brainpower and global R&D. As an organized political force, however, the HTC is relatively new. It was only toward the end of the twentieth century that nongovernmental actors emerged as the dominant funders of R&D in the United States. Moreover, it was only in the 1990s that ICT firms began organizing to shape policies supporting global innovation. The extent and nature of collaboration between the corporate and academic wings of the HTC has also varied. Whether informally or formally, firms and universities have worked most closely on skilled immigration, an area in which their interests are closely aligned. Universities took the lead on sustaining flows of international students after 2001, since they were naturally inclined and well positioned to do so, although firms were interested in the outcome as well. Firms have taken the lead in sustaining openness to global R&D, since the specific challenge that has emerged here (anti-offshoring legislation) has threatened their interests in particular.

The third finding is that the level of openness in U.S. policy broadly reflects the nature of the organized resistance faced by the HTC, as theorized in chapter 2. When the HTC has encountered negligible resistance from other groups—as it did when the foreign student program was challenged following the terrorist attacks of September 11, 2001—it has succeeded in maintaining an uncapped flow of foreign talent. When the HTC has encountered resistance primarily from labor, it has been largely (but not entirely) successful. This was true in the fights over the H-1B cap from 1998 to 2004 and in the battles over anti-offshoring legislation. Last, when the HTC has encountered resistance from large citizen groups, it has failed to achieve its goals. These groups have posed a challenge that the HTC has not been able to overcome, and this is why it has failed to secure increases in skilled immigration since 2004. The HTC is powerful but not unstoppable, and its most formidable opponent thus far has not been labor but citizen groups.

The alternative explanations outlined in chapter 2 all struggle to explain this pattern of U.S. policy. The strategic hegemon hypothesis, with its emphasis on realpolitik relative gains calculations, does not explain the rise and fall of the H-1B cap over time, as explained in chapter 3. It also struggles to explain the extent of U.S. openness to foreign students and to offshoring, as explained in chapters 4 and 5. More generally, the fact that the executive branch has frequently failed to win approval for its preferred policies is difficult for this approach to explain. The Clinton administration was hardly enthusiastic about raising the H-1B cap in 1998, but the cap was raised that year nonetheless. Presidents Bush and Obama were more interested in increasing skilled immigration, but their efforts to do so through comprehensive legislation were unsuccessful. Last, although the Obama administration supported anti-offshoring legislation, these efforts were unsuccessful as well.

Public opinion sheds remarkably little light on U.S. policy. Public attitudes toward skilled immigration appear to have softened from the late 1990s to 2013, but U.S. policy became less open during this time, as explained in chapter 3. In addition, public opinion was hardly an impediment to imposing a moratorium on student visas following the September 11 attacks, as noted in chapter 4, but no such restriction was undertaken. Public opinion has been decidedly negative on offshoring, as explained in chapter 5, but the United States has remained largely open to it.

The preferred party hypothesis finds more support in the case studies. Labor clearly has greater ability to influence Democratic lawmakers than their Republican counterparts, whereas anti-immigration groups are most effective when trying to influence Republicans. Even so, labor has generally struggled to compete with the HTC for influence in Washington, regardless of which political party has been more prominent. This was true even in the late 1990s, when labor attempted to defend the status quo under a Democratic president. Labor's failure to win approval for anti-offshoring legislation in 2010, when the Democrats controlled the White House and both houses of Congress, is also striking. In contrast, large citizen groups have been tough opponents for the HTC on a consistent basis.

The status quo hypothesis also offers some insight into the cases. In their fight against anti-offshoring legislation, high-tech firms were defending the legislative status quo, and this clearly made their task easier. The status quo bias also seems relevant in chapter 4, since the HTC was initially attempting to preserve extant policy toward foreign students. Yet the HTC succeeded not only when attempting to maintain a favorable status quo in 2001, but also when it fought to change an increasingly unfavorable status quo in the years thereafter. This suggests that the status quo bias was less important in this case than the absence of organized opposition to the HTC. In addition, the status quo bias cannot explain variations in U.S. policy toward skilled immigration. The HTC has consistently sought to change the status quo in this area, but its level of success has varied dramatically over time.

The two coalition strength hypotheses struggle to explain U.S. policy, but in different ways. The first predicts that the HTC will enjoy the most success when both its corporate and academic wings are actively engaged. This is poorly supported by the evidence. The HTC was most successful in the case of foreign students, when the academic wing was far more active, and highly successful in the case of offshoring, when only the corporate wing was involved. Evidently, each wing of the HTC is sufficiently powerful to prevail when opposition from other groups is either absent or weak. The second coalition strength hypothesis predicts that the HTC will be most successful when its corporate wing allies with other business interests. There is some support for this view, particularly since high-tech firms' alliance with other business groups helps to explain their relative success in defeating anti-offshoring measures. In other cases, however, high-tech firms have allied with other business groups and failed. This has been evident in the battles over comprehensive immigration reform after 2004, when the HTC allied with a range of other business interests as well as other groups. This powerful coalition was consistently unsuccessful because of the intense opposition it encountered.

What are the wider implications of these findings? In the remainder of this chapter, I approach this question from several different angles. I first consider this study's implications for international relations, and in particular, the study of global innovation. I then focus on the implications of U.S. policies for the United States, and in particular, the country's position vis-à-vis China and India. Finally, I consider the implications of the 2016 U.S. federal elections and how U.S. policies may evolve in the future.

IMPLICATIONS FOR INTERNATIONAL RELATIONS

Several years ago, Robert Keohane charged that scholarship in international political economy had been "remarkably reluctant to focus on major changes taking place in world politics" in recent decades.[1] The globalization of innovation is a new phenomenon particularly deserving of scrutiny. As noted at the outset of this study, global innovation has important implications for the distribution of wealth, the balance of power, and economic interdependence in the twenty-first century. Yet the U.S. policies behind the rise of global innovation have not been subjected to great scrutiny by international relations scholars. Although scholars have begun to explore the politics of skilled immigration or offshoring in the United States, this study has emphasized that these two policy arenas are intertwined as complementary aspects of a larger phenomenon. Furthermore, it has shown that U.S. policies in these arenas can be understood through a common analytical framework focused on a particular set of societal actors: the HTC, labor, and citizen groups. Last, by focusing on U.S. relations with China and India, this study has offered new insight into the political economy of power transitions—a subject that has received far too little scrutiny in the past.

This study also sheds new light on the work that has been done on specific aspects of U.S. policies toward global innovation. Some of the initial work on the politics of offshoring in the United States has focused on public opinion regarding this phenomenon.[2] This is important work, but as I have shown, public opinion has not been the driving force behind U.S. policies in this arena thus far. Recent work on skilled immigration, in turn, has focused on the preferences of skilled natives toward high-skill immigration policies, and scholars have debated whether skilled natives are likely to perceive skilled immigrants as an economic threat.[3] This study supports recent work showing that skilled natives in the same sector may perceive an economic threat from such immigration, even when the economy is relatively strong. Yet this study also demonstrates that the ability of skilled natives to resist increases in skilled immigration in the United States is quite limited.

Whereas previous research has shown effective resistance at the subnational level, this study demonstrates that skilled natives have difficulty influencing national policy, even when their cause is taken up by the national labor movement.[4] Citizen groups, in contrast, can be far more effective.

This study also illuminates a key difference between the politics surrounding high-skill and low-skill immigration in the United States. Scholars have shown that firms' demand for low-skill immigration falls in response to trade openness and capital mobility, with the result that national policy becomes more restrictive.[5] This study shows that the same effect is not at work in high-skill immigration. U.S. high-tech firms have fought to expand skilled immigration in recent decades, notwithstanding substantial trade openness and firm mobility. This persistent interest in skilled immigration presumably reflects the more complex nature of the work involved. Offshoring skilled labor can be considerably more challenging than offshoring low-skilled labor. Moreover, when skilled technical work is to be offshored, firms sometimes find it useful to bring foreign workers to the United States before sending them back overseas; this is why the H-1B is sometimes called the "outsourcing visa." In this way, some kinds of skilled immigration may support offshoring, rather than represent an alternative to it.

Going forward, there is much we still need to learn about the politics of global innovation. While my study has focused on the contention between the HTC and opposing interest groups in the formation of U.S. policies, future research should explore how the HTC fares when it faces resistance from the national executive. In particular, we need more insight into policy outcomes when the HTC's pursuit of openness conflicts with the executive's attempts to ensure national security. It is tempting to imagine that the national political leadership will prevail in any such clash. In reality, such interactions are more likely to resemble negotiations, and multiple outcomes are possible. The interactions between the HTC and the executive branch regarding foreign scholars and students after 2001, as discussed in chapter 4, are noteworthy in this regard. The HTC was generally successful in maintaining an open system, even when the Department of Commerce sought to tighten deemed export controls in 2005. In other cases, however, high-tech interests may well be less successful. Future work should strive to explain such variation.

More generally, future research should consider how other countries grapple with global innovation. The United States stands out in historical terms as a pioneer in this realm. At the outset, the United States played a dominant role in attracting knowledge workers from abroad, and it faced little competition from other developed countries.[6] The United States has also played a key role in pioneering certain aspects of global R&D. U.S. companies, including IBM, Microsoft, and Texas Instruments, have led the way in developing R&D facilities in the developing world and in offshoring more generally.[7] The United States also stands out in international R&D alliances and in international scientific collaboration. Yet the United States is hardly the only developed country that has chosen to participate in global innovation. This raises the question, how have other developed countries responded to this phenomenon?

Recent work has explored how developed countries have adopted new policies toward skilled immigration, rather than policies toward global innovation more generally. Even so, this work suggests that there are several approaches worth pursuing in future research. The first treats states as rational and unitary actors.[8] In this regard, states can be understood as "following the hegemon" in an effort to develop their own high-tech sectors. Some work in this area depicts developed countries as locked in a competition for high-tech talent, with skilled immigration policy a critical tool in this struggle.[9] A second approach emphasizes domestic politics. The work in this vein emphasizes cross-national variation in policy outcomes and argues that this variation reflects domestic institutions and interests.[10] A third approach also stresses variation in policy outcomes but emphasizes the importance of differing ideologies and immigration histories to explain them.[11] Synthetic approaches, combining external and internal forces, are also possible.[12] Future research should continue in all of these veins, but it should consider both policies toward skilled immigration and policies toward global innovation more generally. Future research could also explore the conditions under which each theoretical approach is likely to prove most useful.

A second important avenue for future research concerns the policies of middle-income and developing countries toward global innovation. There is a growing body of research on the kinds of policies that enable "late innovators" to become more important players in global innovation networks. Among other things, this literature has highlighted policies that solve market failures, attract foreign investment, and connect local firms with global markets.[13] My own work has highlighted how late innovators must balance reforms that encourage greater R&D with reforms that limit government supervision of R&D.[14] Less attention has been paid to explaining the varying policy choices of middle-income and developing countries toward global innovation.[15] Moreover, when scholars have focused on policy choice, they have typically considered either policies toward foreign investment or policies toward migration. A comprehensive explanation ought to consider policies in both domains.

IMPLICATIONS FOR THE UNITED STATES, CHINA, AND INDIA

To explain U.S. policies toward global innovation, I have emphasized the lobbying and political battles of particular interest groups. Such groups are a much maligned force in politics. In fact, scholars have highlighted the workings of special interests to explain the decline of major world powers. Mancur Olson's *The Rise and Decline of Nations*, for example, famously posits that stable democracies eventually succumb to "institutional sclerosis," in which the accumulation of special interests diverts policymaking from the "national" interest and reduces economic growth.[16] In reality, the story is unlikely to be quite so simple. Interest groups perform multiple functions in a democracy, including the provision of information to policymakers— even if that information is self-serving. Indeed, some scholars stress that the key question is whether vested interests in a particular country inhibit technological change and stifle innovation, rather than whether they are particularly numerous.[17] In a globalizing world, I would add that we must consider how vested interests and national policies affect the development of other countries as well. Or to use Gilpin's wording, we must consider how vested interests affect the diffusion of "inventiveness." Against this backdrop, it is worth asking, have the interest group battles depicted in this book, and the policies that have arisen from them, accelerated American decline vis-à-vis rising powers such as China and India?

I cannot answer such a wide-ranging and complex question definitively in the space available here. It is possible, however, to make a few broad observations. First, sustained U.S. openness to foreign students has generally worked to the United States' advantage, as far as sustaining innovation leadership is concerned. As noted in chapter 1, students from China and India who earn doctoral degrees in S&E fields from U.S. universities have traditionally been likely to remain in the country for at least five years after graduation.[18] This has enabled a remarkable rise in the share of foreign-born individuals with doctoral degrees in S&E professions in the United States in recent decades, reaching a high of 42 percent in 2013. Even so, we should not assume that the United States will remain quite so attractive in the future. The stay rates for Chinese and Indian students have declined over time, and, as outlined in chapter 1, China in particular has experienced a massive increase in the number of students returning, even if it still struggles to attract some of the brightest. Peking University's chemistry department, to pick one example, includes graduates of the California Institute of Technology (Caltech); Columbia University; Harvard University; the Massachusetts Institute of Technology (MIT); the University of California, Berkeley; the University of Chicago; and the University of Pennsylvania, among other leading U.S. universities. If China's relative attractiveness to its academic diaspora continues to increase, the foreign student program could come to have a more ambiguous effect on the United States' technological leadership in the future.

This brings us to the question of skilled immigration. In this arena, there is clear evidence of dysfunction in the U.S. political system. In a nutshell, the politics that have prevented the United States from reforming its skilled immigration system—whether through a targeted bill or comprehensive legislation—have compromised the ability of the United States to compete for talented S&E workers. The H-1B visa, as a guest-worker program, "is not America's most effective welcome mat," as one of its creators has put it.[19] Even some who advocate expanding the program concede that it suffers from serious problems, including limits on worker mobility, with "deleterious effects."[20] Critics, meanwhile, point to more deficiencies.[21] Even more serious, however, is the U.S. failure to reform the employment-based (EB) visa program. As a means of permanent immigration, the EB visa system is a particularly powerful tool in U.S. efforts to attract talented workers. Yet per-country limits and the massive backlog that has developed, particularly for applicants from China and India, can leave talented workers in limbo for many years. For some applicants, in fact, it has become impossible to estimate how long they will have to wait—or even whether they will ever receive a visa.[22] At some point, talented workers may well conclude they have better options elsewhere. The failure to address this problem, as outlined in chapter 3, is a direct result of the interest-group politics surrounding immigration reform in the United States in the twenty-first century.

The United States' relatively open approach to offshoring is more difficult to assess. Most U.S. business R&D spending still takes place in the United States, and most U.S. R&D spending abroad takes place in Europe rather than Asia.[23] And although U.S. firms' R&D in China and India is growing, the impact it has on the local innovation capabilities of these countries is difficult to assess. Recent research indicates that foreign multinationals' R&D in China has become much more productive over time.[24] How much this development helps or hurts Chinese organizations remains unclear, however. One study of Chinese firms found that increasing foreign R&D actually had a negative effect on technical change in those firms.[25] Yet other studies have offered more positive assessments.[26] The impact in India is also hard to gauge. A study published in 2010 found that foreign R&D that seeks to create new technology has a positive effect on domestic firms, whereas that which exploits existing technology for the local market has a negative effect.[27] It is worth asking whether this is still the case as foreign R&D in India has grown, and also whether the same effect can now be seen in China. In short, the overall impact of foreign R&D in China and India remains unclear.

Going forward, the impact of global R&D on China and India will depend on a range of variables. It will certainly depend on the kinds of economic reforms adopted in China and India in years to come. These could either strengthen or constrain the ability of Chinese and Indian firms to profit from global R&D, as I have argued elsewhere.[28] But it will also depend on the approach taken by the U.S. government and U.S. firms. Bills designed to limit U.S. firms' overseas investments are unlikely to pass muster in Congress, as the track record demonstrates. A more productive approach would entail greater collaboration between the public and private sectors to protect the intellectual property in multinational networks. Some such efforts are already underway, but there is more that could be done. In 2007, two prominent officials in the Department of Commerce suggested that new thinking was needed in terms of how the U.S. government approaches export control, and they suggested that companies should consider drawing up custom risk-mitigation plans for government consideration, based on the specific technology and the intended end user.[29] The point was not to create a new export-control regime, but to help companies take greater ownership of their long-term vulnerabilities while also addressing the increasing complexity of technology control in the twenty-first century.[30] Although such customized discussions have yet to become commonplace, a few companies have had this kind of dialogue with the U.S. government regarding their operations in China.[31] Looking ahead, this kind of cooperative approach may face challenges of its own, but it is more likely to prove feasible than restrictions on overseas R&D.[32]

In short, the evidence is mixed. There are signs of dysfunction in the U.S. approach to global innovation in some regards but not in others. In some respects, the effect of U.S. policies is difficult to assess and could change in the future. Indeed, this is a very brief and provisional assessment, and more research is clearly needed. We should not conclude, however, that the problems in U.S. skilled immigration policy represent a form of Olsonian "sclerosis" from which the United States cannot escape. Interest groups, including firms, labor unions, and anti-immigration forces, have sparred over U.S. immigration policy since the nineteenth century, and U.S. policies have fluctuated greatly over time.[33] Moreover, the strength of different groups, particularly citizen groups, can wax and wane as popular enthusiasm for particular causes fluctuates. With that in mind, let us turn to the future of U.S. policymaking toward global innovation.

LOOKING AHEAD

The election of Donald Trump as president of the United States in November 2016 shattered much conventional wisdom about the U.S. political system. Trump's campaign was noteworthy for its anti-immigrant, anti-globalization, and anti-minority themes, even engaging in what Speaker of the House Paul Ryan called "textbook" racism.[34] Trump's victory seemed to catapult the U.S. anti-immigration movement into the Oval Office, while marking a resounding political defeat for the HTC, which had largely supported Hillary Clinton.[35] As of May 2017, the full implications of Trump's election for U.S. policymaking remain unclear. Many vacancies in the executive branch remain unfilled, and given the controversy surrounding Trump's relationship with Russia, among other issues, there is speculation as to how long he will remain in office. Even so, it is worth asking, is U.S. policymaking shifting to a new phase in which the state eclipses societal interest groups, particularly the HTC, in policy toward global innovation?

This is unlikely, for several reasons. First, the most significant changes to U.S. policies in this realm generally require the White House to work with Congress. Indeed, it is for this reason that the Obama administration attempted to revise U.S. policies toward offshoring and immigration through legislation. The need for congressional action, in turn, creates considerable opportunities for powerful interests to block changes they would find unwelcome. High-tech firms, along with broader business interests, have successfully stymied most attempts to revise U.S. policy toward offshoring for more than a decade, even when such changes have been supported by the White House. Citizen groups have blocked several attempts to change U.S. immigration policy over the same time period. It remains possible that the Trump administration will manage to advance some of its agenda through legislation, particularly while the GOP controls both houses of Congress. Major legislation that disregards the collective interests of the HTC, however, is unlikely.

The executive branch can act unilaterally, of course, but the potential for such action faces constraints. Although the Obama administration did make some changes to skilled immigration policy through executive action, for example, it avoided making broader changes not clearly available under the law, as noted in chapter 3. Indeed, Obama's changes left the executive director of Compete America fuming that many "actions" and "concrete outcomes" were still needed.[36] The Trump administration could act with a more expansive notion of executive power, but it also faces the threat of legal action if it does so. In the first half of 2017, for example, both high-tech firms and universities actively supported legal challenges to Trump's efforts to impose a travel ban on certain Muslim-majority countries, deeply frustrating the administration.[37] However this dispute is ultimately resolved, it highlights that the executive branch does not possess carte blanche to remake U.S. policy as it sees fit, even if it can tarnish the U.S. image by attempting to do so.

The second key point is that the relationship between the Trump White House and the HTC is more complex and can be more cooperative than is sometimes apparent. To be sure, there is much conflict between the two. The dispute over the travel ban is an obvious case in point. The Trump administration also frustrated technology firms early on by making changes to the H-1B visa program, notably suspending expedited application processing and barring entry-level computer work from being considered a "specialty occupation."[38] Trump could also undo some of Obama's regulatory changes, including work authorization for H-1B spouses and the Optional Practical Training (OPT) extension for science, technology, engineering, and mathematics (STEM) graduates. In the meantime, media reports claim Silicon Valley is working to "topple" Trump, including through the work of new groups that support Democratic candidates, such as Tech for Campaigns.[39] The Trump–tech conflict is intense, even visceral, and it could escalate in the future.

Yet there is also a dialogue occurring between the Trump administration and the HTC, and there are some areas in which they could cooperate. In April 2017, the White House launched a review of the H-1B visa program designed to ensure that visas "are awarded to the most-skilled or highest-paid petition beneficiaries."[40] As of this writing, it remains unclear where this review will lead or whether the administration will pursue legislation toward this end. It is possible, however, that the review will generate proposals welcomed by a substantial part of the HTC. Depending on the details, many leading U.S. ICT firms could benefit from a more merit-based system, since they typically hire more educated H-1B workers and pay higher salaries than other firms do. Whether such a system would also benefit U.S. universities would depend on the particulars; an emphasis on skills could benefit recent graduates, but a new salary threshold or other requirements might not. In contrast, a more merit-based system would likely reduce the visas available to IT outsourcing firms, including not only Indian firms but also U.S. firms (such as IBM) that engage in this business. This would affect not only the IT outsourcing firms themselves but also their customers, of whom many are leading U.S. ICT firms.[41] U.S. firms of all kinds could also be concerned about the reaction to these changes abroad, particularly in India. On balance, some members of the U.S. high-tech community may support a more merit-based system, whereas others resist such a change.

Tax reform represents a clear arena of potential cooperation between the Trump administration and U.S. high-tech firms. As a candidate, Trump railed against the offshoring of U.S. jobs and threatened firms with some kind of border tax to punish such behavior. The border tax idea has proven controversial within the Trump administration, however, and has not been a focus of its tax reform efforts thus far.[42] Instead, the administration has focused on spurring corporate investment in the United States, particularly by lowering the corporate income tax rate and offering a tax holiday to encourage multinationals to repatriate profits from abroad.[43] These ideas naturally appeal to high-tech firms.[44] Even so, the administration will be disappointed if it expects a tax holiday to spur new high-tech investments in the United States. The largest ICT firms are not capital constrained, and repatriated profits would more likely be used to reduce debt, fund share-holder dividends, and finance stock repurchases.[45]

The final point concerns staying power. Although Trump's election marked a stunning change of course for the United States, it did not erase the basic societal interests that have driven U.S. policy over the past two decades. High-tech interests will remain a potent political force long after Trump has left the White House, and they will continue to press for greater openness to global innovation. The extent to which the HTC succeeds will be powerfully shaped by the nature of the resistance it encounters. This resistance could become more formidable in the future, particularly if the anti-immigration movement intensifies after Trump leaves office. Alternatively, the resistance could become weaker. The anti-immigration movement could atrophy over time, or groups resisting piecemeal reforms could relent in some cases or find themselves sidelined. If so, the United States could become more open to collaboration with China and India in the future. Whether the United States will pursue such collaboration in a more intelligent way, one that addresses the shortcomings of its current approach, remains to be seen.

NOTES

INTRODUCTION

1. On how the feature was created, I am indebted to a personal communication from Joy Ann Lo, senior communications manager at Microsoft Research Asia (from 2011 to 2014), September 22, 2015.
2. National Science Foundation, *Science and Engineering Indicators 2016* (Arlington, VA: National Science Foundation, 2016), chapter 3, 101.
3. National Science Foundation, *Science and Engineering Indicators 2016*, chapter 2, 71.
4. IBM, "IBM Research: Global Labs," *IBM*, April 3, 2015, www.research.ibm.com/labs/; Sujit John, "Cisco Needs to Align with Indian Government's Goals," *Times of India*, July 2, 2014.
5. James D. Fearon, "Rationalist Explanations for War," *International Organization* 49, no. 3 (1995): 405; Dustin H. Tingley, "The Dark Side of the Future: An Experimental Test of Commitment Problems in Bargaining," *International Studies Quarterly* 55, no. 2 (June 2011): 521–44.
6. Susan K. Sell, *Private Power, Public Law: The Globalization of Intellectual Property Rights* (Cambridge: Cambridge University Press, 2003); Jonathan D. Aronson, "International Intellectual Property Rights in a Networked World," in *Power, Interdependence, and Nonstate Actors in World Politics*, ed. Helen V. Milner and Andrew Moravcsik (Princeton, NJ: Princeton University Press, 2011), 185–203.
7. Vivek Wadhwa et al., *Losing the World's Best and Brightest: America's New Immigrant Entrepreneurs, Part V* (Kansas City, MO: Ewing Marion Kauffman Foundation, March 2009). Evidence regarding reverse migration is discussed in chapter 1.
8. Colin Powell, "Remarks at the Elliott School of International Affairs," (speech, George Washington University, Washington, DC, September 5, 2003), https://2001-2009.state .gov/secretary/former/powell/remarks/2003/23836.htm.

9. Gary P. Freeman and David K. Hill, "Disaggregating Immigration Policy: The Politics of Skilled Labor Recruitment in the U.S.," *Knowledge, Technology & Policy* 19, no. 3 (2006): 7–26; Lucie Cerna, "The Varieties of High-Skilled Immigration Policies: Coalitions and Policy Outputs in Advanced Industrial Countries," *Journal of European Public Policy* 16, no. 1 (January 2009): 144–61; Lucie Cerna, "Attracting High-Skilled Immigrants: Policies in Comparative Perspective," *International Migration* 52, no. 3 (June 2014): 69–84; Monica Boyd, "Recruiting High Skill Labour in North America: Policies, Outcomes and Futures," *International Migration* 52, no. 3 (June 2014): 40–54; Chris F. Wright, "Why Do States Adopt Liberal Immigration Policies? The Policymaking Dynamics of Skilled Visa Reform in Australia," *Journal of Ethnic and Migration Studies* 41, no. 2 (January 28, 2015): 306–28.

10. Edward D. Mansfield and Diana C. Mutz, "US Versus Them: Mass Attitudes Toward Offshore Outsourcing," *World Politics* 65, no. 4 (2013): 571–608; Kerry A. Chase, "Moving Hollywood Abroad: Divided Labor Markets and the New Politics of Trade in Services," *International Organization* 62, no. 4 (2008): 653–87.

11. Mark Zachary Taylor, *The Politics of Innovation: Why Some Countries Are Better Than Others at Science and Technology* (Oxford: Oxford University Press, 2016); Joel W. Simmons, *The Politics of Technological Progress* (Cambridge: Cambridge University Press, 2016); Linda Weiss, *America Inc.? Innovation and Enterprise in the National Security State* (Ithaca, NY: Cornell University Press, 2014); Joseph Wong, *Betting on Biotech: Innovation and the Limits of Asia's Developmental State* (Ithaca, NY: Cornell University Press, 2011); Dan Breznitz, *Innovation and the State* (New Haven, CT: Yale University Press, 2007).

12. In this book, the term *high tech* is typically used to refer to ICT industries in particular. The concept of the high-tech community is explained in chapter 3.

13. On the challenge of approximating a controlled comparison in qualitative research, see Alexander George and Andrew Bennett, *Case Studies and Theory Development in the Social Sciences* (Cambridge, MA: MIT Press, 2005), 151–79.

14. Some of these interviews were conducted purely on a background basis. Others are cited in the chapters that follow.

15. Elhanan Helpman, *The Mystery of Economic Growth* (Cambridge, MA: Harvard University Press, 2004).

16. Christina L. Davis and Sophie Meunier, "Business as Usual? Economic Responses to Political Tensions," *American Journal of Political Science* 55, no. 3 (July 2011): 628–46.

17. Edward D. Mansfield and Brian Pollins, "Interdependence and Conflict: An Introduction," in *Economic Interdependence and International Conflict: New Perspectives on an Enduring Debate*, ed. Edward D. Mansfield and Brian Pollins (Ann Arbor: University of Michigan Press, 2003); John Ravenhill, "The Economics-Security Nexus in the Asia-Pacific region," in *Security Politics in the Asia-Pacific: A Regional-Global Nexus?*, ed. William Tow (New York: Cambridge University Press, 2009), 188–207.

18. Stephen G. Brooks, *Producing Security: Multinational Corporations, Globalization, and the Changing Calculus of Conflict* (Princeton, NJ: Princeton University Press, 2005).

19. Dale C. Copeland, *Economic Interdependence and War* (Princeton, NJ: Princeton University Press, 2014); Dale C. Copeland, "Economic Interdependence and War: A Theory of Trade Expectations," *International Security* 20, no. 4 (1996): 5–41.

20. Robert Gilpin, *U.S. Power and the Multinational Corporation: The Political Economy of Foreign Direct Investment* (New York: Basic Books, 1975), 67; Robert Gilpin, *War and Change in World Politics* (Cambridge: Cambridge University Press, 1983), 182.

21. William R. Thompson, "Long Waves, Technological Innovation, and Relative Decline," *International Organization* 44, no. 2 (1990): 201–33; George Modelski and William R. Thompson, *Leading Sectors and World Powers: The Coevolution of Global Politics and Economics* (Columbia: University of South Carolina Press, 1996).

22. Tai Ming Cheung and Bates Gill, "Trade Versus Security: How Countries Balance Technology Transfers with China," *Journal of East Asian Studies* 13, no. 3 (2013): 445.

1. THE RISE OF GLOBAL INNOVATION

1. Andrew B. Kennedy, "Slouching Tiger, Roaring Dragon: Comparing India and China as Late Innovators," *Review of International Political Economy* 23, no. 2 (2016): 1–28. On the debate over China's and India's trajectories in innovation more generally, see Andrew B. Kennedy, "Powerhouses or Pretenders? Debating China's and India's Emergence as Technological Powers," *The Pacific Review* 28, no. 2 (2015): 281–302.

2. For a helpful overview of the concept of innovation, see Jan Fagerberg, "Innovation: A Guide to the Literature," in *The Oxford Handbook of Innovation*, ed. Jan Fagerberg, David C. Mowery, and Richard R. Nelson (Oxford: Oxford University Press, 2005), 4–9.

3. "New-to-the-world" technologies are often contrasted with those that are "new to the country" or "new to the firm." See Organisation for Economic Co-operation and Development (OECD), *The Measurement of Scientific and Technological Activities (Oslo Manual)* (Paris: OECD, 1997).

4. Fagerberg, "Innovation," 4–9.

5. In addition, when a given technology consists of components that combine in a modular fashion, it is common to distinguish between two additional types of innovation. "Modular" innovation refers to changes in individual components, whereas "architectural" innovation combines previously existing components in new ways. See Dieter Ernst, *A New Geography of Knowledge in the Electronics Industry? Asia's Role in Global Innovation Networks* (Honolulu, HI: East-West Center, 2009), 7–12; Richard M. Henderson and Kim B. Clark, "Architectural Innovation: The Reconfiguration of Existing Product Technologies and the Failure of Established Firms," *Administrative Science Quarterly* 35, no. 1 (March 1990): 9–30.

6. Daniele Archibugi and Simona Iammarino, "The Globalization of Technological Innovation: Definition and Evidence," *Review of International Political Economy* 9, no. 1 (Spring 2002): 98–122; Rajneesh Narula and Antonello Zanfei, "Globalization of Innovation: The Role of Multinational Enterprises," in *The Oxford Handbook of Innovation*,

ed. Jan Fagerberg, David C. Mowery, and Richard R. Nelson (Oxford: Oxford University Press, 2005), 318–45; Dieter Ernst, *Innovation Offshoring: Exploring Asia's Emerging Role in Global Innovation Networks*, East-West Center Special Report No. 10 (Honolulu, HI: East-West Center, July 2006).

7. In this sense, the approach taken here resonates with that taken in William Lazonick, *Sustainable Prosperity in the New Economy? Business Organization and High-Tech Employment in the United States* (Kalamazoo, MI: W. E. Upjohn Institute, 2009).

8. This is not to suggest that innovation is simply a function of increasing inputs of labor and capital. The context in which capital and labor operate matters a great deal, as the literature on national and regional innovation systems makes clear. See Christopher Freeman, *Technology, Policy, and Economic Performance: Lessons from Japan* (London: Pinter, 1987); Bengt-Åke Lundvall, *National Systems of Innovation: Toward a Theory of Innovation and Interactive Learning* (London: Pinter, 1992); Richard R. Nelson, ed., *National Innovation Systems: A Comparative Analysis* (Oxford: Oxford University Press, 1993); Bjørn T. Asheim and Meric S. Gertler, "The Geography of Innovation: Regional Innovation Systems," in *The Oxford Handbook of Innovation*, ed. Jan Fagerberg, David C. Mowery, and Richard R. Nelson (Oxford: Oxford University Press, 2005), 291–317. An intriguing line of inquiry is to explore how globalization is changing national innovation systems. See David M. Hart, "Understanding Immigration in a National Systems of Innovation Framework," *Science & Public Policy* 34, no. 1 (2007): 45–53.

9. Paul M. Romer, "Endogenous Technological Change," *Journal of Political Economy* 98, no. 5 (1990): S71–S102.

10. An additional 3 percent were self-employed. See National Science Foundation, *Science and Engineering Indicators 2016* (Arlington, VA: National Science Foundation), chapter 3, 37.

11. National Science Foundation, *Science and Engineering Indicators 2016*, chapter 3, 40.

12. National Science Foundation, *Science and Engineering Indicators 2016*, chapter 5, appendix table 18.

13. National Science Foundation, *Science and Engineering Indicators 2016*, chapter 5, 76–78.

14. National Science Foundation, *Science and Engineering Indicators 2016*, chapter 5, 76.

15. Keith Pavitt, "Innovation Processes," in *The Oxford Handbook of Innovation*, ed. Jan Fagerberg, David Mowery, and Richard R. Nelson (Oxford: Oxford University Press, 2006), 95; David C. Mowery and Bhaven N. Sampat, "Universities in National Innovation Systems," in *The Oxford Handbook of Innovation*, ed. Jan Fagerberg, David Mowery, and Richard R. Nelson (Oxford: Oxford University Press, 2006), 221–24.

16. Barry R. Chiswick and Timothy Hatton, "International Migration and the Integration of Labor Markets," in *Globalization in Historical Perspective*, ed. Michael D. Bordo, Alan M. Taylor, and Jeffrey G. Williamson (Chicago: University of Chicago Press, 2003), 70.

17. Harm G. Schröter and Anthony S. Travis, "An Issue of Different Mentalities: National Approaches to the Development of the Chemical Industry in Britain and Germany Before 1914," in *The Chemical Industry in Europe, 1850–1914*, ed. Ernst Homburg, Anthony S. Travis, and Harm G. Schröter (Dordrecht, Netherlands: Kluwer, 1998), 100.

18. Petra Moser, Alessandra Voena, and Fabian Waldinger, "German Jewish Émigrés and US Invention," *The American Economic Review* 104, no. 10 (2014): 3222–55; Annie Jacobsen, *Operation Paperclip: The Secret Intelligence Program That Brought Nazi Scientists to America* (New York: Little, Brown, 2014).

19. The term *brain drain* was first used to describe the flow of British scientists to the United States in the early 1960s but soon became associated with flows from developing countries to the developed world. See Hart, "Understanding Immigration in a National Systems of Innovation Framework," 46.

20. Frédéric Docquier, Olivier Lohest, and Abdeslam Marfouk, "Brain Drain in Developing Countries," *The World Bank Economic Review* 21, no. 2 (January 1, 2007): 194.

21. Cansin Arslan et al., *A New Profile of Migrants in the Aftermath of the Recent Economic Crisis*, OECD Social, Employment and Migration Working Paper No. 160 (2014): 37, www.oecd.org/els/mig/WP160.pdf.

22. Author's calculations based on Arslan et al., *A New Profile of Migrants*, 26.

23. AnnaLee Saxenian, *The New Argonauts: Regional Advantage in a Global Economy* (Cambridge, MA: Harvard University Press, 2006).

24. National Science Foundation, *Science and Engineering Indicators 2016*, chapter 3, 103.

25. Institute for Regional Studies, "Silicon Valley Index 2017," *Silicon Valley Indicators*, February 2017, 14, http://jointventure.org/images/stories/pdf/index2017.pdf.

26. OECD, *Education at a Glance 2014: OECD Indicators* (Paris: OECD, 2014), 344.

27. OECD, *Education at a Glance 2014*, 342.

28. OECD, *Education at a Glance 2014*, 350–51.

29. National Science Foundation, *Science and Engineering Indicators 2016*, chapter 2, 91.

30. Australian Government Department of Education and Training, "International Students Studying Science, Technology, Engineering and Mathematics (STEM) in Australian Higher Education Institutions," *Research Snapshots*, October 2015, https://international education.gov.au/research/Research-Snapshots/Documents/STEM%202014.pdf.

31. National Science Foundation, *Science and Engineering Indicators 2016*, chapter 2, appendix table 43.

32. National Science Foundation, *Science and Engineering Indicators 2004* (Arlington, VA: National Science Foundation, 2004), chapter 2, appendix table 27; National Science Foundation, *Science and Engineering Indicators 2016*, chapter 2, 71.

33. The number of Chinese and Indian S&E graduate students in the United States is discussed in some detail in chapter 4.

34. National Science Foundation, *Science and Engineering Indicators 2016*, chapter 5, 81.

35. National Science Foundation, *Science and Engineering Indicators 2016*, chapter 2, appendix table 33.

36. National Science Foundation, *Science and Engineering Indicators 2016*, chapter 5, appendix table 16.

37. National Science Foundation, *Science and Engineering Indicators 2016*, chapter 3, 105.

38. The rise of industrial R&D is discussed at greater length in chapter 2.

39. National Science Foundation, *Science and Engineering Indicators 2012* (Arlington, VA: National Science Foundation, 2012), chapter 4, 15 and 55.

40. These two categories are roughly based on previous work by Archibugi and Iammarino, though I treat flows of students separately, as noted earlier. See Archibugi and Iammarino, "The Globalization of Technological Innovation."

41. Steve Fraser, "The Hollowing Out of America," *The Nation*, December 3, 2012, www .thenation.com/article/171563/hollowing-out-america#.

42. National Science Foundation, *Science and Engineering Indicators 2014* (Arlington, VA: National Science Foundation, 2014), chapter 4, 27.

43. Narula and Zanfei, "Globalization of Innovation," 326.

44. On the way in which multinationals contend with weak intellectual property protection in particular, see Minyuan Zhao, "Conducting R&D in Countries with Weak Intellectual Property Rights Protection," *Management Science* 52 (August 2006): 1185–99.

45. Bureau of Economic Analysis, "Foreign Direct Investment in the U.S., Majority-Owned Bank and Nonbank U.S. Affiliates, Research and Development Expenditures for 2013," *International Data: Direct Investment and Multinational Enterprises*, 2016, www.bea.gov /iTable/index_MNC.cfm.

46. Bureau of Economic Analysis, "U.S. Direct Investment Abroad, All Majority-Owned Foreign Affiliates, Research and Development Expenditures for 2013," *International Data: Direct Investment and Multinational Enterprises*, 2016, www.bea.gov/iTable/index_MNC.cfm.

47. Andrew B. Kennedy, "Unequal Partners: U.S. Collaboration with China and India in Research and Development," *Political Science Quarterly* 132, no. 1 (2017): 63–86.

48. Bureau of Economic Analysis, "U.S. Direct Investment Abroad." The "information" category includes software publishers. Data for software publishers were not available in 2013, but in previous years, this subcategory dominated overseas R&D spending in the information category.

49. For a discussion of these two types, see Narula and Zanfei, "Globalization of Innovation," 326–29.

50. In recent years, scholars have noted additional types of overseas R&D. For example, see Jian Wang, Lan Xue, and Zheng Liang, "Multinational R&D in China: From Home-Country-Based to Host-Country-Based," *Innovation: Management, Policy & Practice* 14, no. 2 (June 2012): 192–202.

51. Melissa Schilling, "Technology Shocks, Technological Collaboration, and Innovation Outcomes," *Organization Science* 26, no. 3 (May–June 2015): 668–86.

52. Stephen G. Brooks, *Producing Security: Multinational Corporations, Globalization, and the Changing Calculus of Conflict* (Princeton, NJ: Princeton University Press, 2005), 32–33.

53. Note that no single database contains every single alliance created, although there is broad agreement among them as to the varying number of alliances over time. Figures based on the SDC Platinum database should thus be seen as suggestive rather than definitive. See Melissa Schilling, "Understanding the Alliance Data," *Strategic Management Journal* 30, no. 3 (2009): 233–60.

54. Ernst and Young, *Globalizing Venture Capital: Global Venture Capital Insights and Trends Report*, 2012, 7, www.ey.com/Publication/vwLUAssets/Globalizing_venture_capital_VC _insights_and_trends_report_CY0227/$FILE/Globalizing%20venture%20capital _VC%20insights%20and%20trends%20report_CY0227.pdf.

55. KPMG and CB Insights, *Venture Pulse 2016* (New York: CB Insights, April 13, 2016), 55, https://www.cbinsights.com/research-venture-capital-Q1-2016.

56. KPMG and CB Insights, *Venture Pulse 2016*, 16.

57. "About SMART," *Singapore–MIT Alliance for Research and Technology*, 2013, http://smart .mit.edu/about-smart/about-smart.html.

58. Peter H. Koehn, "Developments in Transnational Research Linkages: Evidence from US Higher-Education Activity," *Journal of New Approaches in Educational Research* 3, no. 2 (2014): 53.

59. Jason Lane and Kevin Kinser, "Is Today's University the New Multinational Corporation?" *The Conversation*, June 5, 2015, http://theconversation.com/is-todays -university-the-new-multinational-corporation-40681.

60. National Science Foundation, *Science and Engineering Indicators 2016*, chapter 5, appendix table 41.

61. On China in particular, see Denis Fred Simon and Cong Cao, *China's Emerging Techno- logical Edge: Assessing the Role of High-End Talent* (New York: Cambridge University Press, 2009), 238.

62. "Navigating China's Tech Jungle," *Business Times*, September 1, 2012.

63. "Guruduth Banavar," *LinkedIn*, accessed February 11, 2016, www.linkedin.com/in/banavar.

64. Prasad Ram, founder and chief executive officer, Gooru, interview by author, Palo Alto, California, July 15, 2014.

65. China's prominence in these visa programs is discussed in chapter 4.

66. National Science Foundation, *Science and Engineering Indicators 2016*, chapter 5, appendix table 56.

67. Andrew B. Kennedy, *The International Ambitions of Mao and Nehru: National Efficacy Beliefs and the Making of Foreign Policy* (New York: Cambridge University Press, 2012); Andrew B. Kennedy, "India's Nuclear Odyssey: Implicit Umbrellas, Diplomatic Disappointments, and the Bomb," *International Security* 36, no. 2 (2011): 120–53.

68. On the danger global R&D can pose for developing countries, see Ernst, *A New Geography of Knowledge?*, 38–39.

69. State Council of the People's Republic of China, "Guojia Zhongchangqi Kexue He Jishu Fazhan Guihua Gangyao (2006–2020 Nian) [National Medium- and Long-Term Pro- gram for Science and Technology Development (2006–2020)]," part 2, section 2, *Zhongguo Zhengfu Menhu Wangzhan [Chinese Government Gateway Website]*, February 9, 2006, www .gov.cn/jrzg/2006-02/09/content_183787.htm.

70. William C. Hannas, James Mulvenon, and Anna B. Puglisi, *Chinese Industrial Espionage: Technology Acquisition and Military Modernization* (New York: Routledge, 2013); Kennedy, "Slouching Tiger, Roaring Dragon."

71. David Zweig and Changgui Chen, *China's Brain Drain to the United States: Views of Overseas Chinese Students and Scholars in the 1990s* (Berkeley: Institute of East Asian Studies, University of California, 1995), 19.

72. David M. Lampton, *A Relationship Restored: Trends in U.S.–China Educational Exchanges, 1978–1984* (Washington, DC: National Academies, 1986), 23.

73. Leo A. Orleans, "China's Changing Attitude Toward the Brain Drain and Policy Toward Returning Students," *China Exchange News* 17, no. 2 (1989): 2.

74. David Zweig, "Learning to Compete: China's Efforts to Encourage a 'Reverse Brain Drain,'" in *Competing for Global Talent*, ed. Christiane Kuptsch and Eng Fong Pang (Geneva: International Labor Office, 2006), 189.

75. Orleans, "China's Changing Attitude," 2.

76. Foreign Broadcast Information Service, "Foreign Schooling Policy Remains Unchanged," *FBIS-CHI-88-224*, November 14, 1988, 31.

77. Guojia Jiaowei Shehui Kexue Yanjiu yu Yishu Jiaoyusi, Guojia Jiaowei Sixiang Zhengzhi Gongzuosi, and Beijing Shiwei Gaodeng Xuexiao Gongzuo Weiyuanhui, *Wushitian de Huigu Yu Fansi [Looking Back and Reflecting on Fifty Days]* (Beijing: Gaodeng Jiaoyu Chubanshe, 1989), 72–82.

78. Heping Shi, "Beijing's China Card," *Harper's Magazine*, September 1990, 28.

79. Deng Xiaoping, "Gist of Speeches Made in Wuchang, Shenzhen, Zhuhai and Shanghai (From January 18 to February 21, 1992)," *Beijing Review*, February 7, 1994, 16.

80. Paul Englesberg, "Reversing China's Brain Drain: The Study-Abroad Policy, 1978–1993," in *Great Policies: Strategic Innovations in Asia and the Pacific Basin*, ed. John Dickey Montgomery and Dennis A. Rondinelli (Westport, CT: Praeger, 1995), 117.

81. Communist Party of China, "Zhonggong Zhongyang Guanyu Jianli Shehuizhuyi Shichang Jingji Tizhi Ruogan Wenti de Jueding [Chinese Communist Party Decision Regarding Several Questions in Building a Socialist Market Economy System]," *Zhongguo Gongchandang Xinwen Wang [Chinese Communist Party News Network]*, November 14, 1993, http://cpc.people.com.cn/GB/64162/134902/8092314.html.

82. Simon and Cao, *China's Emerging Technological Edge*, 219.

83. Ellis Rubinstein, "China's Leader Commits to Global Science and Scientific Exchange," *Science*, October 6, 2000, www.sciencemag.org/careers/2000/10/chinas-leader-commits-global-science-and-scientific-exchange. For more on Jiang's approach, see David Zweig and Huiyao Wang, "Can China Bring Back the Best? The Communist Party Organizes China's Search for Talent," *The China Quarterly* 215 (2013): 594.

84. United Nations Educational, Scientific and Cultural Organization (UNESCO), "Education Data," *UIS.Stat*, 2014, www.uis.unesco.org/datacentre/pages/default.aspx.

85. Zweig, "Learning to Compete," 194.

86. Zweig and Wang, "Can China Bring Back the Best?," 596.

87. "Qianren Jihua [Thousand Talents Plan]," *Qianren Jihua Wang*, 2016, www.1000plan.org /qrjh/section/2.

88. Hong Liu and Els van Dongen, "China's Diaspora Policies as a New Mode of Transnational Governance," *Journal of Contemporary China* 25, no. 102 (2016): 15.

89. Huiyao Wang, David Zweig, and Xiaohua Lin, "Returnee Entrepreneurs: Impact on China's Globalization Process," *Journal of Contemporary China* 20, no. 70 (June 2011): 415.

90. Charlotte Liu et al., "Turning Point: Chinese Science in Transition" (Shanghai: Nature Publishing Group, November 2015), 8, www.nature.com/press_releases/turning_point.pdf; Chi Ma, "Famous Science Projects Face Axe in Funding Overhaul," *China Daily*, January 8, 2015, www.chinadaily.com.cn/china/2015-01/08/content_19275863.htm.

91. Yu Wei and Zhaojun Sun, "China: Building an Innovation Talent Program System and Facing Global Competition in a Knowledge Economy," *The Academic Executive Brief*, 2012, http://academicexecutives.elsevier.com/articles/china-building-innovation-talent -program-system-and-facing-global-competition-knowledge.

92. "Guanyu Gongbu Diyipi Qingnian Qianren Jihua Yinjin Rencai Mingdan de Gonggao [Proclamation Announcing the First Batch of Names for the Thousand Youth Talents Plan to Attract Talent]," *Qianren Jihua Wang*, November 11, 2011, http://1000plan.org/qrjh /article/18053.

93. "Dishiyipi Qianren Jihua Qingnian Rencai, Chuangye Rencai Ruxuan Mingdan [Name List for the Eleventh Batch of Awardees Under the Thousand Youth Talents and Startup Talents Plan]," *Qianren Jihua Wang*, May 13, 2015, http://1000plan.org/qrjh/article/61716; "Guanyu Gongbu Dishier Pi Guojia Qianren Jihua Qingnian Rencai, Chuanye Rencai Ruxuan Renyuan Mingdan de Gonggao [Proclamation Announcing the Twelfth Batch of Awardees for the Thousand Youth Talents and Startup Talents Plan]," *Qianren Jihua Wang*, March 14, 2016, http://1000plan.org/qrjh/article/61716; "Guanyu Gongbu Dishisan Pi Guojia Qianren Jihua Qingnian Xiangmu Chuangye Rencai Xiangmu Ruxuan Renyuan Mingdan de Gonggao [Announcement of the Thirteenth Batch of Awardees for the Youth and Startup Talent Programs of the National Thousand Talents Plan]," *Qianren Jihua Wang*, May 11, 2017, http://1000plan.org/qrjh/article/69239.

94. "ORISE Workforce Studies Infographics—StayRates," *Oak Ridge Institute for Science and Education*, March 31, 2017, https://public.tableau.com/views/ORISEWorkforceStudies Infographics-StayRates-mobilefriendly/5-YearStayRates?%3Aembed=y&%3AshowViz Home=no&%3Adisplay_count=y&%3Adisplay_static_image=y&%3AbootstrapWhen Notified=true.

95. "Ruxuan Zhongguo Qianren Jihua Waizhuan Xiangmu de Zhuanjia Da 381 Ming [Three Hundred Eighty-One Individuals Selected for China's Thousand Talents Foreign Experts Program]," *Kexue Wang [Science Net]*, April 15, 2017, http://news.sciencenet.cn/htmlnews /2017/4/373557.shtm.

96. Zweig and Wang, "Can China Bring Back the Best?," 613.

97. Cong Cao et al., "Reforming China's S&T System," *Science* 341, no. 6145 (2013): 462.

98. Quoted in Zweig and Wang, "Can China Bring Back the Best?," 606.

99. Jacques Gaillard and Anne-Marie Gaillard, "Introduction: The International Mobility of Brains—Exodus or Circulation?" *Science, Technology & Society* 2, no. 2 (1997): 195–228.

100. Zweig, "Learning to Compete," 192–93.

101. David Zweig, Chung Siu Fung, and Donglin Han, "Redefining the Brain Drain: China's 'Diaspora Option,'" *Science, Technology & Society* 13, no. 1 (2008): 1–33.

102. Weihua Chen, "US Immigration Reform a Challenge for China," *China Daily*, February 8, 2013, http://europe.chinadaily.com.cn/opinion/2013-02/08/content_16216633.htm.

103. Quoted in Ruowei Cheng, "Woguo Liushi Dingjian Rencai Shu Ju Shijie Shouwei [China's Loss of Top Talent Is Greatest in the World]," *Renmin Ribao [People's Daily]*, June 6, 2013, http://finance.people.com.cn/n/2013/0606/c1004-21754321.html.

104. Huiyao Wang, *Rencai Zhanzheng [Talent War]* (Beijing: China CITIC, 2009).

105. "Guancha: Zhongguo Gaoduan Rencai Liushi Lu Jugao Buxia [Observation: The Rate of China's High-End Brain Drain Remains High]," *Zhongguo Jiaoyu Bao [China Education Daily]*, January 20, 2014, http://edu.sina.com.cn/a/2014-01-20/1148238838.shtml.

106. "Gonggu Fazhan Zui Guangfan de Aiguo Tongyi Zhanxian Wei Shixian Zhongguo Meng Tigong Guangfan Liliang Zhichi [Consolidate and Develop the Most Wide-Ranging Patriotic United Front to Provide Extensive Supporting Strength to the Chinese Dream]," *Renmin Ribao [People's Daily]*, May 21, 2015, http://paper.people.com.cn/rmrb /html/2015-05/21/nw.D110000renmrb_20150521_2-01.htm.

107. China Scholarship Council, "2012 Nian Guojia Liuxue Jijin Zizhu Chuguo Liuxue Renyuan Xuanpai Jianzhang [Briefing on the Selection of Awardees for Study Abroad Financial Support in 2012]," November 1, 2011, http://v.csc.edu.cn/Chuguo/739b1b8c118441e 5bb211c388563f7da.shtml; China Scholarship Council, "2017 Nian Guojia Liuxue Jijin Zizhu Chuguo Liuxue Renyuan Xuanpai Jianzhang [Briefing on the Selection of Awardees for Study Abroad Financial Support in 2017]," December 12, 2016, www.csc.edu.cn/article/709.

108. China Scholarship Council, "2015 Nian Guojia Gongpai Chuguo Liuxue Xuanpai Jihua Queding [Government Determines Plan for Study Abroad Awards in 2015]" (Beijing: Chinese Education Online and Best Choice for Education, October 30, 2014), www.csc .edu.cn/News/2acf973ba1a84ca69f5386a574771906.shtml.

109. Quoted in Robert Buderi and Gregory T. Huang, *Guanxi (The Art of Relationships): Microsoft, China, and Bill Gates' Plan to Win the Road Ahead* (London: Random House, 2006), 57–58.

110. Dan Breznitz and Michael Murphree, *Run of the Red Queen: Government, Innovation, Globalization, and Economic Growth in China* (New Haven, CT: Yale University Press, 2011), 106.

111. Kennedy, "Slouching Tiger, Roaring Dragon," 71–72.

112. Colum Murphy and Lilian Lin, "For China's Jobseekers, Multinational Companies Lose Their Magic," *Wall Street Journal*, April 3, 2014, http://blogs.wsj.com/chinarealtime /2014/04/03/for-chinas-jobseekers-multinational-companies-lose-their-magic/?mod =chinablog. I am also indebted to an interview with a Chinese government official on this point. Interview by author, Beijing, June 25, 2013.

113. State Council of the People's Republic of China, "Guojia Zhongchangqi," part 8, section 8.

114. State Council of the People's Republic of China, "Guowuyuan Guanyu Jiakuai Peiyu He Fazhan Zhanluexing Xinxing Chanye de Jueding [State Council Decision on Accelerating the Cultivation and Development of Strategic Emerging Industries]," *Zhongguo Zhengfu Menhu Wangzhan [Chinese Government Gateway Website]*, October 10, 2010, part 6, www .gov.cn/zwgk/2010-10/18/content_1724848.htm.

115. State Council of the People's Republic of China, "Guomin Jingji He Shehui Fazhan Dishierge Wunian Guihua Gangyao (Quan Wen) [Compendium of the Twelfth Five-Year Plan for Development of the National Economy and Society (Full Text)]," part 20, chapter 52, section 1, *Zhongguo Zhengfu Menhu Wangzhan* [*Chinese Government Gateway Website*], March 16, 2011, www.gov.cn/2011lh/content_1825838.htm.

116. Nick Marro, "Foreign Company R&D: In China, For China," *China Business Review*, June 1, 2015.

117. Marro, "Foreign Company R&D."

118. Matthew Miller, "Spy Scandal Weighs on U.S. Tech Firms in China, Cisco Takes Hit," *Reuters*, November 14, 2013, www.reuters.com/article/2013/11/14/us-china-cisco-idUSBRE9 AD0J420131114.

119. People's Republic of China Ministry of Science and Technology official, interview by author, Beijing, July 3, 2013.

120. "China to Overtake US as New Frontier for Global R&D," *People's Daily Online*, January 27, 2014, http://english.peopledaily.com.cn/98649/8523078.html.

121. Google China employee, interview by author, July 3, 2013.

122. Gert Bruche, "The Emergence of China and India as New Competitors in MNCs' Innovation Networks," *Competition & Change* 13, no. 3 (2009): 276; Edward S. Steinfeld, *Playing Our Game: Why China's Rise Doesn't Threaten the West* (New York: Oxford University Press, 2010), 152.

123. Steinfeld, *Playing Our Game*, 152; Loren Brandt and Eric Thun, "The Fight for the Middle: Upgrading, Competition, and Industrial Development in China," *World Development* 38, no. 11 (November 2010): 1566–69.

124. Interview by author, Beijing, July 4, 2013. My interlocutor in this case wished to remain anonymous.

125. Lee Branstetter, Guangwei Lee, and Francisco Veloso, "The Rise of International Coin-vention," in *The Changing Frontier: Rethinking Science and Innovation Policy*, ed. Adam B. Jaffe and Benjamin F. Jones (Chicago: University of Chicago Press, 2015), 162.

126. Sylvia Schwaag-Serger, "Foreign Corporate R&D in China: Trends and Policy Issues," in *The New Asian Innovation Dynamics: China and India in Perspective*, ed. Govindan Parayil and Anthony P. D'Costa (New York: Palgrave Macmillan, 2009), 50–78.

127. Joy Ann Lo, senior communications manager, Microsoft Research Asia, interview by author, February 19, 2014.

128. "R&D Technology Center China," *GE Lighting Asia Pacific*, accessed April 16, 2014, www .gereveal.ca/LightingWeb/apac/resources/world-of-ge-lighting/research-and-development /china-technology-centre.jsp.

129. "Intel China Research Center," *Intel*, accessed May 6, 2014, www.intel.com/cd/corporate /icrc/apac/eng/about/167066.htm.

130. Kennedy, "Unequal Partners," 78–79.

131. State Council of the People's Republic of China, "Guojia Zhongchangqi."

132. American Enterprise Institute, "China Global Investment Tracker," *AEI*, April 30, 2017, www.aei.org/china-global-investment-tracker/.

133. Louise Lucas, "US Concerns Grow Over Chinese Chip Expansion," *Financial Times*, January 16, 2017, https://www.ft.com/content/fb2e4454-c36e-11e6-9bca-2b93a6856354.

134. State Council of the People's Republic of China, "Guojia Zhongchangqi," part 8, section 8.

135. State Council of the People's Republic of China, "Guowuyuan Guanyu Jiakuai Peiyu He Fazhan Zhanluexing Xinxing Chanye de Jueding," part 4.

136. Xiaolan Fu and Hongru Xiong, "Open Innovation in China: Policies and Practices," *Journal of Science and Technology Policy in China* 2, no. 3 (2011): 204–205, 207–208.

137. Thomson Reuters, SDC Platinum database, accessed April 27, 2015, access via subscription only.

138. Douglas B. Fuller, *Paper Tigers, Hidden Dragons: Firms and the Political Economy of China's Technological Development* (Oxford: Oxford University Press, 2016), 60; Di Guo and Kun Jiang, "Venture Capital Investment and the Performance of Entrepreneurial Firms: Evidence from China," *Journal of Corporate Finance* 22 (2013): 377.

139. Fuller, *Paper Tigers, Hidden Dragons*, 61.

140. Mark Humphery-Jenner and Jo-Ann Suchard, "Foreign Venture Capitalists and the Internationalization of Entrepreneurial Companies: Evidence from China," *Journal of International Business Studies* 44, no. 6 (2013): 607–21.

141. "China Encourages More VC Funding, Promises Foreign Firms Equal Treatment," *Reuters*, September 1, 2016, www.reuters.com/article/us-china-economy-idUSKCN1175BT.

142. Government of India Department of Science and Technology, *Science, Technology and Innovation Policy 2013* (New Delhi: Ministry of Science and Technology, 2013), 4, www.dst.gov.in/sti-policy-eng.pdf.

143. Narendra Modi, "PM's Speech to 104th Session of the Indian Science Congress, Tirupati (Full Text)," *Microfinance Monitor*, January 3, 2017, www.microfinancemonitor.com/pms-speech-to-104th-session-of-the-indian-science-congress-tirupati-full-text/43799.

144. Devesh Kapur, *Diaspora, Development, and Democracy: The Domestic Impact of International Migration from India* (Princeton, NJ: Princeton University Press, 2010), 51–54.

145. Government of India, Passports Act of 1967, Pub. L. No. 15 (1967), 1, http://passportindia.gov.in/AppOnlineProject/pdf/passports_act.pdf.

146. Government of India, Passports Act of 1967, 6.

147. Government of India, Emigration Act of 1983, Pub. L. No. 31 (1983), 15, http://moia.gov.in/writereaddata/pdf/emig_act.pdf.

148. David Fitzgerald, "Inside the Sending State: The Politics of Mexican Emigration Control," *International Migration Review* 40, no. 2 (2006): 262.

149. The main exception lies in the medical sphere, in which some restrictions have been introduced but with too many loopholes to be effective. See Binod Khadria, "Brain Drain, Brain Gain, India," in *The Encyclopedia of Global Human Migration*, ed. Immanuel Ness (New York: Wiley-Blackwell, 2013), 743.

150. J. Singh and V. V. Krishna, "Trends in Brain Drain, Gain and Circulation: Indian Experience of Knowledge Workers," *Science Technology & Society* 20, no. 3 (November 1, 2015): 302–6.

151. Kapur, *Diaspora, Development, and Democracy*, 180–81.

152. Rajiv Gandhi, "Revamping the Educational System," *Indian National Congress*, August 29, 1985, www.inc.in/resources/speeches/345-Revamping-the-Educational-System.

153. "Vajpayee Calls for Reversing Brain Drain, Cutting Red Tape," *Hindu Business Line*, January 4, 2003, www.thehindubusinessline.com/bline/2003/01/04/stories/2003010402410500 .htm; "PM for Reverse Brain Drain of Scientists," *Economic Times*, January 4, 2011, http://articles.economictimes.indiatimes.com/2011-01-04/news/28424740_1 _scientists-of-indian-origin-talent-pool-98th-indian-science.

154. Kapur, *Diaspora, Development, and Democracy*, 165. The emigration rate is the number of emigrating individuals in a given educational group divided by the total number of individuals in that group within the sending country.

155. On the previous three sentences, see Sanjoy Chakravorty, Devesh Kapur, and Nirvikar Singh, *The Other One Percent: Indians in America* (Oxford: Oxford University Press, 2016), x, 30, 108.

156. UNESCO, "Education Data"; Devesh Kapur, "Indian Higher Education," in *American Universities in a Global Market*, ed. Charles T. Clotfelter (Chicago: University of Chicago Press, 2010), 326.

157. Institute for International Education, "International Student Totals by Place of Origin, 2013/14–2014/15," *Open Doors Report on International Educational Exchange*, 2015, www .iie.org/Research-and-Publications/Open-Doors/Data/International-Students /All-Places-of-Origin/2013-15.

158. National Science Foundation, *Science and Engineering Indicators 2016*, chapter 2, appendix table 27.

159. Australian Government Department of Education, "International Student Enrolments by Nationality in 2013," April 2014, 1, https://internationaleducation.gov.au/research /research-snapshots/pages/default.aspx; UNESCO, "Global Flow of Tertiary-Level Students," *UIS.Stat*, accessed December 10, 2014, www.uis.unesco.org/EDUCATION /Pages/international-student-flow-viz.aspx.

160. Rajiv Gandhi, "Keep Pace with Technology," *Indian National Congress*, December 19, 1985, www.inc.in/resources/speeches/315-Keep-Pace-with-Technology.

161. Singh and Krishna, "Trends in Brain Drain," 308–10.

162. In 2001, India introduced the Model Education Loan Scheme to provide soft loans to students pursuing higher education, whether in India or abroad. However, the scheme does not require students to return to India, and it has been criticized as cumbersome and lacking in reach. See Asian Development Bank, *Counting the Cost: Financing Asian Higher Education for Inclusive Growth* (Manila: Asian Development Bank, 2012), 16.

163. Kennedy, "Slouching Tiger, Roaring Dragon," 71–72.

164. Chakravorty, Kapur, and Singh, *The Other One Percent*, 56.

165. Chakravorty, Kapur, and Singh, *The Other One Percent*, 56; Vivek Wadhwa, *The Immigrant Exodus: Why America Is Losing the Global Race to Capture Entrepreneurial Talent* (Philadelphia: Wharton Digital, 2012); Singh and Krishna, "Trends in Brain Drain," 310–13.

166. Michael G. Finn, "Stay Rates of Foreign Doctorate Recipients from U.S. Universities, 2001" (Oak Ridge, TN: Oak Ridge Institute for Science and Education, November 2003), 9, http://orise.orau.gov/files/sep/stay-rates-foreign-doctorate-recipients-2001.pdf.

167. Michael G. Finn, "Stay Rates of Foreign Doctorate Recipients from U.S. Universities, 2011" (Oak Ridge, TN: Oak Ridge Institute for Science and Education, January 2014), 7, http://orise.orau.gov/science-education/difference/stay-rates-impact.aspx.

168. Fei Qin, "Global Talent, Local Careers: Circular Migration of Top Indian Engineers and Professionals," *Research Policy* 44, no. 2 (2015): 405–20.

169. Khadria, "Brain Drain, Brain Gain, India," 3.

170. High Level Committee on the Indian Diaspora, "The Indian Diaspora" (New Delhi: Ministry of External Affairs, 2001), xi, http://indiandiaspora.nic.in/contents.htm.

171. For a helpful overview of India's diaspora policies, see Daniel Naujoks, *Migration, Citizenship, and Development: Diasporic Membership Policies and Overseas Indians in the United States* (New Delhi: Oxford University Press, 2013), 49–65.

172. Kapur, *Diaspora, Development, and Democracy*, 261–68.

173. Narendra Modi, "Modi Speaks in San Jose: The Indian Prime Minister in His Own Words," *Silicon Valley One World*, September 27, 2015, www.siliconvalleyoneworld.com/2015/09/30/modi-speaks-in-san-jose-the-indian-prime-minister-in-his-own-words/.

174. Kapur, *Diaspora, Development, and Democracy*, 106–13.

175. World Bank Migration and Remittances Team, *Migration and Remittances: Recent Developments and Outlook*, Migration and Development Brief No. 23 (October 6, 2014): 4, 10, http://siteresources.worldbank.org/INTPROSPECTS/Resources/334934-1288990760745/MigrationandDevelopmentBrief23.pdf.

176. Ronil Hira, "U.S. Immigration Regulations and India's Information Technology Industry," *Technological Forecasting and Social Change* 71, no. 8 (October 2004): 837–54.

177. That same year, six outsourcing firms headquartered elsewhere (but which also employ many mobile Indian workers) received 9,943 H-1B visas. See Haeyoun Park, "How Outsourcing Companies Are Gaming the Visa System," *New York Times*, November 10, 2015.

178. Nair Chendakera, "U.S., India Move to Boost Cooperation in R&D," *Electronic Engineering Times*, March 27, 2000; Bruce Stokes, "India's Paradox," *National Journal*, April 7, 2007; Amiti Sen, "India to Ask US for More H-1B Visas," *Economic Times*, October 19, 2009; Nirupama Rao, "America Needs More High-Skilled Worker Visas," *USA Today*, April 14, 2013, www.usatoday.com/story/opinion/2013/04/14/india-trade-technology-column/2075159/.

179. "India May Drag US to WTO for Hiking H-1B Visa Fee," *Times of India*, August 17, 2010, http://timesofindia.indiatimes.com/business/india-business/India-may-drag-US-to-WTO-for-hiking-H-1B-visa-fee/articleshow/6325497.cms.

180. "Obama Assures Modi on Concerns Over H-1B Visa Issue," *Times of India*, January 26, 2015, http://timesofindia.indiatimes.com/india/Obama-assures-Modi-on-concerns-over-H-1B-visa-issue/articleshow/46022377.cms.

181. Ronil Hira, "New Data Show How Firms Like Infosys and Tata Abuse the H-1B Program," *Economic Policy Institute*, February 19, 2015, www.epi.org/blog/new-data-infosys-tata-abuse-h-1b-program/.

182. Arun Janardhanan, "US Move to Lure Science Grads Worries India," *Times of India—Chennai Edition*, February 4, 2013.

183. Former high-ranking Indian official, interview by author, February 16, 2017, New Delhi. My interlocutor noted his belief that individuals who wished to remain abroad would ultimately find a way to do so, regardless of visa caps.

184. Dinesh C. Sharma, *The Long Revolution: The Birth and Growth of India's IT Industry* (Noida: Harper Collins, 2009), 317–18.

185. "Research and Development," *General Electric*, 2017, www.ge.com/in/about-us/research-and-development.

186. Min Ye, *Diasporas and Foreign Direct Investment in China and India* (New York: Cambridge University Press, 2014), 177–204.

187. N. Mrinalini, Pradosh Nath, and G. D. Sandhya, "Foreign Direct Investment in R&D in India," *Current Science* 105, no. 6 (September 2013): 770–71.

188. Sujit John and Shilpa Phadnis, "For MNCs, India Still an R&D Hub and It's Growing," *Times of India*, March 2, 2017, http://timesofindia.indiatimes.com/city/bengaluru/for-mncs-india-still-an-rd-hub-and-its-growing/articleshow/57421665.cms.

189. Bruche, "The Emergence of China and India as New Competitors in MNCs' Innovation Networks," 276.

190. IBM India Research Lab, Bangalore, interview by author, January 13, 2014. My interlocutor in this case wished to remain anonymous.

191. Nirmalya Kumar and Phanish Puranam, *India Inside: The Emerging Innovation Challenge to the West* (Cambridge, MA: Harvard Business Press, 2012), 109–14.

192. Government of India, "Science, Technology and Innovation Policy," 15.

193. Thomson Reuters, SDC Platinum database.

194. Narendra Modi, "PM's Address to the Nation from the Ramparts of the Red Fort on the Sixty-Eighth Independence Day," *PMIndia*, August 15, 2014, http://pmindia.gov.in/en/news_updates/text-of-pms-address-in-hindi-to-the-nation-from-the-ramparts-of-the-red-fort-on-the-68th-independence-day/.

195. Narendra Modi, "Text of PM Shri Narendra Modi's Address at the 102nd Indian Science Congress," *Narendra Modi*, January 3, 2015, www.narendramodi.in/text-of-pm-shri-narendra-modis-address-at-the-102nd-indian-science-congress.

196. Narendra Modi, "Modi Speaks in San Jose: The Indian Prime Minister in His Own Words," *SiliconValleyOneWorld*, September 30, 2015, www.siliconvalleyoneworld.com/2015/09/30/modi-speaks-in-san-jose-the-indian-prime-minister-in-his-own-words/.

197. On this and the following two sentences, see Rafiq Dossani and Martin Kenney, "Creating an Environment for Venture Capital in India," *World Development* 30, no. 2 (2002): 243–49.

198. Samarth Agarwal, interview by author, February 16, 2017. Agarwal is the founder of Spaceboat, an Indian startup firm established in 2016.

199. "One Year of Startup India: Report Card," *TechCircle*, January 16, 2017.

200. "RBI Eases Norms for Foreign Investment in Startups," *TechCircle*, October 21, 2016.

2. INNOVATION LEADERSHIP AND CONTESTED OPENNESS

1. For an earlier review, see G. John Ikenberry, David A. Lake, and Michael Mastanduno, "Introduction: Approaches to Explaining American Foreign Economic Policy," in *The State and American Foreign Economic Policy*, ed. G. John Ikenberry and David A. Lake (Ithaca, NY: Cornell University Press, 1988), 1–14.

2. Charles Poor Kindleberger, *The World in Depression, 1929–1939* (Berkeley: University of California Press, 1973); Stephen D. Krasner, "State Power and the Structure of International Trade," *World Politics* 28, no. 3 (1976): 322; David A. Lake, "Leadership, Hegemony, and the International Economy: Naked Emperor or Tattered Monarch with Potential?" *International Studies Quarterly* 37, no. 4 (December 1993): 459. See also G. John Ikenberry, *After Victory: Institutions, Strategic Restraint, and the Rebuilding of Order After Major Wars* (Princeton, NJ: Princeton University Press, 2001); Rafael Reuveny and William R. Thompson, *Growth, Trade, and Systemic Leadership* (Ann Arbor: University of Michigan Press, 2009); John Ravenhill, "US Economic Relations with East Asia: From Hegemony to Complex Interdependence?," in *Bush and Asia: America's Evolving Relations with East Asia*, ed. Mark Beeson (London: Routledge, 2006), 43–45.

3. Stephan Haggard, "The Institutional Foundations of Hegemony: Explaining the Reciprocal Trade Agreements Act of 1934," *International Organization* 42, no. 1 (1988): 91–119; Michael Mastanduno, *Economic Containment: CoCom and the Politics of East–West Trade* (Ithaca, NY: Cornell University Press, 1992); Hugo Meijer, *Trading with the Enemy: The Making of US Export Control Policy Toward the People's Republic of China* (Oxford: Oxford University Press, 2016).

4. Judith Goldstein, *Ideas, Interests, and American Trade Policy* (Ithaca, NY: Cornell University Press, 1993); James Ashley Morrison, "Before Hegemony: Adam Smith, American Independence, and the Origins of the First Era of Globalization," *International Organization* 66, no. 3 (2012): 395–428.

5. For a few examples, see Helen V. Milner, *Resisting Protectionism: Global Industries and the Politics of International Trade* (Princeton, NJ: Princeton University Press, 1988), 103–58; Susan K. Sell, *Private Power, Public Law: The Globalization of Intellectual Property Rights* (Cambridge: Cambridge University Press, 2003), 75–120; Llewelyn Hughes, *Globalizing Oil: Firms and Oil Market Governance in France, Japan, and the United States* (New York: Cambridge University Press, 2014), 149–97; Margaret E. Peters, "Trade, Foreign Direct Investment, and Immigration Policy Making in the United States," *International Organization* 68, no. 4 (2014): 811–44; In Song Kim, "Political Cleavages Within Industry: Firm Level Lobbying for Trade Liberalization," *American Political Science Review* 111 (2017): 1–20.

6. For example, see I. M. Destler, *American Trade Politics* (New York: Columbia University Press, 2005).

7. Joseph A. Schumpeter, *Business Cycles: A Theoretical, Historical, and Statistical Analysis of the Capitalist Process* (New York: McGraw-Hill, 1939), 72–192.

8. Robert Gilpin, *U.S. Power and the Multinational Corporation: The Political Economy of Foreign Direct Investment* (New York: Basic Books, 1975), 69.

9. Robert Gilpin, *War and Change in World Politics* (Cambridge: Cambridge University Press, 1983), 182. See also Robert Gilpin, *Global Political Economy: Understanding the International Economic Order* (Princeton, NJ: Princeton University Press, 2001), 140–41.

10. William R. Thompson, "Long Waves, Technological Innovation, and Relative Decline," *International Organization* 44, no. 2 (1990): 201–33; George Modelski and William R. Thompson, *Leading Sectors and World Powers: The Coevolution of Global Politics and Economics* (Columbia: University of South Carolina Press, 1996).

11. Modelski and Thompson, *Leading Sectors and World Powers*, 70–71.

12. Daniel Drezner, "State Structure, Technological Leadership and the Maintenance of Hegemony," *Review of International Studies* 27, no. 1 (2001): 3–25; Espen Moe, "Mancur Olson and Structural Economic Change: Vested Interests and the Industrial Rise and Fall of the Great Powers," *Review of International Political Economy* 16, no. 2 (June 26, 2009): 202–30; Ashley J. Tellis et al., *Measuring National Power in the Post-Industrial Age* (Santa Monica, CA: RAND, 2000), 36–40. Other scholars have noted the importance of innovation in supporting national power more generally. See Mark Zachary Taylor, *The Politics of Innovation: Why Some Countries Are Better Than Others at Science and Technology* (Oxford: Oxford University Press, 2016).

13. Reuveny and Thompson, *Growth, Trade, and Systemic Leadership*, 13.

14. William R. Thompson, "Systemic Leadership, Evolutionary Processes, and International Relations Theory: The Unipolarity Question," *International Studies Review* 8, no. 1 (2006): 4; Tellis et al., *Measuring National Power in the Post-Industrial Age*, 40.

15. ICT industries are typically defined as including both ICT manufacturers and ICT service providers (including software, telecommunications, data processing and hosting, and computer systems design firms). See Brandon Shackelford and John Jankowski, "Information and Communications Technology Industries Account for $133 Billion of Business R&D Performance in the United States in 2013," *National Center for Science and Engineering Statistics*, April 2016, https://www.nsf.gov/statistics/2016/nsf16309/nsf16309.pdf.

16. Organisation for Economic Co-operation and Development (OECD), "Population," *OECD. Stat*, March 23, 2016, http://stats.oecd.org/Index.aspx?DatasetCode=POP_FIVE_HIST.

17. European Commission, "The 2016 EU Industrial R&D Investment Scoreboard," *Economics of Industrial Research and Innovation*, accessed March 13, 2017, http://iri.jrc.ec.europa.eu /scoreboard16.html.

18. The four countries are Finland, Israel, South Korea, and Taiwan. See Shackelford and Jankowski, "Information and Communications Technology Industries Account for $133 Billion of Business R&D Performance in the United States in 2013."

19. Times Higher Education, "World University Rankings 2014–2015," 2014, *Times Higher Education*, www.timeshighereducation.co.uk/world-university-rankings/2014-15/world-ranking.

20. Elhanan Helpman, *The Mystery of Economic Growth* (Cambridge, MA: Harvard University Press, 2004).

21. David S. Landes, *The Unbound Prometheus: Technological Change and Industrial Development in Western Europe from 1750 to the Present* (Cambridge: Cambridge University Press, 2003), 47–50.

22. Paul Kennedy, *The Rise and Fall of the Great Powers* (New York: Vintage, 1987), 151; François Crouzet, *The Victorian Economy* (London: Methuen, 1982), 4–5.

23. Nicholas Bloom, Raffaella Sadun, and John Van Reenen, "Americans Do IT Better: US Multinationals and the Productivity Miracle," *The American Economic Review* 102, no. 1 (2012): 167–201.

24. Barry Jaruzelski, Volker Staack, and Aritomo Shinozaki, "Software-as-a-Catalyst," *Strategy+Business*, October 25, 2016, www.strategy-business.com/feature/Software-as-a-Catalyst?gko=7a1ae.

25. Michael C. Horowitz, *The Diffusion of Military Power: Causes and Consequences for International Politics* (Princeton, NJ: Princeton University Press, 2010).

26. Helen Milner and David B. Yoffie, "Between Free Trade and Protectionism: Strategic Trade Policy and a Theory of Corporate Trade Demands," *International Organization* 43, no. 2 (Spring 1989): 244; Dennis C. Mueller, "First-Mover Advantages and Path Dependence," *International Journal of Industrial Organization* 15, no. 6 (October 1997): 827–50.

27. This paragraph draws on Joel Mokyr, *The Lever of Riches: Technological Creativity and Economic Progress* (New York: Oxford University Press, 1990), 88, 117–19.

28. The following four sentences draw on Johann Peter Murmann and Ralph Landau, "On the Making of Competitive Advantage: The Development of the Chemical Industry in Britain and Germany Since 1850," in *Chemicals and Long-Term Economic Growth: Insights from the Chemical Industry*, ed. Ashish Arora, Ralph Landau, and Nathan Rosenberg (New York: Wiley, 1998), 31–34.

29. David A. Hounshell and John Kenly Smith, *Science and Corporate Strategy: Du Pont R&D, 1902–1980* (Cambridge: Cambridge University Press, 1988), 11.

30. Murmann and Landau, "On the Making of Competitive Advantage," 36–37.

31. Robert Friedel and Paul B. Israel, *Edison's Electric Light: The Art of Invention* (Baltimore, MD: Johns Hopkins University Press, 2010), 1.

32. "General Electric Research Lab History," *Edison Tech Center*, 2015, www.edisontechcenter.org/GEresearchLab.html.

33. The following paragraph draws on David C. Mowery and Nathan Rosenberg, "The U.S. National Innovation System," in *National Innovation Systems: A Comparative Analysis*, ed. Richard R. Nelson (Oxford: Oxford University Press, 1993), 32–40.

34. Mowery and Rosenberg, "The U.S. National Innovation System," 48.

35. Mowery and Rosenberg, "The U.S. National Innovation System," 34.

36. Tony Borroz, "Chevrolet's Mouse That Roared," *Wired*, August 22, 2011, www.wired.com/2011/08/chevrolets-mouse-that-roared/.

37. "G.M. Earnings in '55 Go Over Billion Mark," *Chicago Tribune*, February 3, 1956; Steven Mufson, "Once a Recession Remedy, GM's Empire Falls," *Washington Post*, June 2, 2009; "Fortune 500: 1955 Full List," *Fortune*, 2015, http://archive.fortune.com/magazines/fortune/fortune500_archive/full/1955/.

38. T. A. Heppenheimer, *Turbulent Skies: The History of Commercial Aviation* (New York: Wiley, 1995), 162–69, 183.

39. National Science Foundation, *Science and Engineering Indicators 2012* (Arlington, VA: National Science Foundation, 2012), chapter 4, 13.

40. Mowery and Rosenberg, "The U.S. National Innovation System," 47.

41. Linda Weiss, *America Inc.? Innovation and Enterprise in the National Security State* (Ithaca, NY: Cornell University Press, 2014), 78.

42. For an overview of this transition, see William Lazonick, *Sustainable Prosperity in the New Economy? Business Organization and High-Tech Employment in the United States* (Kalamazoo, MI: W. E. Upjohn Institute, 2009), 1–113.

43. National Science Foundation, *Science and Engineering Indicators 2012*, chapter 4, appendix table 7.

44. Compared with that of its ICT peers, Apple's R&D intensity (R&D expenditure divided by total revenue) has traditionally been relatively low. In recent years, however, the firm's revenues have been so massive that devoting even a small share to R&D has placed Apple among the world leaders in this regard.

45. Barry Jaruzelski, Volker Staack, and Aritomo Shinozaki, "2016 Global Innovation 1000 Study," *PwC*, accessed March 15, 2017, www.strategyand.pwc.com/innovation1000. Note that Amazon is sometimes classified as a "general retailer" and sometimes as an Internet firm. PwC classifies it under "software and Internet."

46. U.S. National Science Foundation, *Science and Engineering Indicators 2016* (Arlington, VA: National Science Foundation), chapter 4, appendix table 7.

47. Weiss, *America Inc.?*, 97.

48. Jacques S. Gansler and William Lucyshyn, *Commercial-Off-the-Shelf (COTS): Doing It Right* (College Park: University of Maryland Center for Public Policy and Private Enterprise, 2008), 55, www.dtic.mil/dtic/tr/fulltext/u2/a494143.pdf.

49. Jessi Hempel, "DOD Head Ashton Carter Enlists Silicon Valley to Transform the Military," *Wired*, November 18, 2015, www.wired.com/2015/11/secretary-of-defense-ashton-carter/.

50. Weiss, *America Inc.?*, 64–121.

51. For 2016, the Obama administration proposed spending $86.4 billion on information technology purchases, of which $37.3 billion was allocated for defense agencies. See John K. Higgins, "Proposed 2016 Federal Budget Plumps IT Spending by $2B," *E-Commerce Times*, March 11, 2015, www.ecommercetimes.com/story/81805.html.

52. Doug Cameron and Alistair Barr, "Google Snubs Robotics Rivals, Pentagon," *Wall Street Journal*, March 5, 2015, www.wsj.com/articles/google-snubs-robotics-rivals-pentagon -1425580734.

53. Olivia Solon, "US Tech Firms Bypassing Pentagon to Protect Deals with China, Strategist Says," *The Guardian*, March 2, 2016, www.theguardian.com/technology/2016/mar/02 /us-tech-firms-pentagon-national-security-china-deals.

54. Except as noted, the following paragraph draws on David M. Hart, "Political Representation in Concentrated Industries: Revisiting the 'Olsonian Hypothesis,'" *Business and Politics* 5, no. 3 (2003): 261–86.

55. For an overview of IBM's relationship with the federal government, see David Hart, "Red, White, and 'Big Blue': IBM and the Business–Government Interface in the United States, 1956–2000," *Enterprise and Society* 8, no. 1 (2007): 1–34.

56. Tony Romm, "Apple Takes Washington," *Politico*, August 27, 2015, http://politi.co /1Px6AWo.

57. *Soft money* refers to contributions to national political parties. Such contributions were first publicly disclosed in the 1991/92 election cycle and were banned by legislation following the 2002 elections. *Outside spending* refers to political expenditures made by groups or individuals independently of, and not in coordination with, candidates' committees.

58. For the data referred to in this paragraph, see Center for Responsive Politics, "Interest Groups," *OpenSecrets.org*, accessed March 16, 2017, www.opensecrets.org/industries/. Note that the data on party preference are based solely on contributions to candidates and parties.

59. For a helpful overview of high-tech industry associations, see David Hart, "New Economy, Old Politics: The Evolving Role of the High-Technology Industry in Washington, D.C.," in *Governance Amid Bigger, Better Markets*, ed. Joseph S. Nye and John D. Donahue (Washington, DC: Brookings Institution, 2004), 247–50.

60. Michael A. Murray, "Defining the Higher Education Lobby," *The Journal of Higher Education* 47, no. 1 (January 1976): 82–83.

61. Washington Higher Education Secretariat, "About WHES," accessed February 3, 2016, www.whes.org/index.html.

62. For the data referred to in this paragraph, see Center for Responsive Politics, "Lobbying Spending Database," *OpenSecrets.org*, accessed March 3, 2017, www.opensecrets.org /lobby/.

63. David M. Hart, "'Business' Is Not an Interest Group: On the Study of Companies in American National Politics," *Annual Review of Political Science* 7 (2004): 47–69; Kim, "Political Cleavages Within Industry."

64. David Hart, "High-Tech Learns to Play the Washington Game, or the Political Education of Bill Gates and Other Nerds," in *Interest Group Politics*, ed. Allan J. Cigler and Burdett Loomis, 6th ed. (Washington, DC: CQ, 2002), 293–312.

65. Lucie Cerna, "The Varieties of High-Skilled Immigration Policies: Coalitions and Policy Outputs in Advanced Industrial Countries," *Journal of European Public Policy* 16, no. 1 (January 2009): 144–61.

66. Shackelford and Jankowski, "Information and Communications Technology Industries Account for $133 Billion of Business R&D Performance in the United States in 2013," 5.

67. For example, see Philip G. Altbach and Jane Knight, "The Internationalization of Higher Education: Motivations and Realities," *Journal of Studies in International Education* 11, no. 3–4 (2007): 292–93.

68. Lesleyanne Hawthorne, "The Growing Global Demand for Students as Skilled Migrants" (Washington, DC: Migration Policy Institute, 2008), 5–7, www.migrationpolicy.org/sites /default/files/publications/intlstudents_0.pdf.

69. The interest of capital in the world's lead economy in a liberal international investment regime has long been recognized. See Gilpin, *U.S. Power and the Multinational Corporation*, 44–78.

70. Rajneesh Narula and Antonello Zanfei, "Globalization of Innovation: The Role of Multinational Enterprises," in *The Oxford Handbook of Innovation*, ed. Jan Fagerberg, David C. Mowery, and Richard R. Nelson (Oxford: Oxford University Press, 2005), 325–26; U.S. National Science Foundation, *Science and Engineering Indicators 2016*, chapter 4, 61–67.

71. Narula and Zanfei, "Globalization of Innovation: The Role of Multinational Enterprises," 328.

72. Gilpin, *U.S. Power and the Multinational Corporation*; Milner, *Resisting Protectionism*; Sell, *Private Power, Public Law*; Peters, "Trade, Foreign Direct Investment, and Immigration Policy Making in the United States."

73. "Fortune 500," *Fortune*, 2015, http://fortune.com/fortune500/.

74. Philip Elmer-DeWitt, "Apple as the Goose That Laid the Golden Eggs. Five of Them," *Fortune*, November 25, 2013, http://fortune.com/2013/11/25/apple-as-the-goose-that-laid -the-golden-eggs-five-of-them/.

75. Meijer, *Trading with the Enemy*, 145–64.

76. On the value of information in lobbying, see Frank R. Baumgartner et al., *Lobbying and Policy Change: Who Wins, Who Loses, and Why* (Chicago: University of Chicago Press, 2009), 123.

77. David Vogel, *Fluctuating Fortunes: The Political Power of Business in America* (New York: Basic Books, 1989); Mark A. Smith, *American Business and Political Power: Public Opinion, Elections, and Democracy* (Chicago: University of Chicago Press, 2000); Sheldon Kamieniecki, *Corporate America and Environmental Policy: How Often Does Business Get Its Way?* (Stanford, CT: Stanford University Press, 2006); Mark A. Smith, "The Mobilization and Influence of Business Interests," in *The Oxford Handbook of American Political Parties and Interest Groups*, ed. L. Sandy Maisel and Jeffrey M. Berry (Oxford: Oxford University Press, 2010), 451–67.

78. Baumgartner et al., *Lobbying and Policy Change*, 202–4.

79. Smith, *American Business and Political Power*, 8–11.

80. Baumgartner et al., *Lobbying and Policy Change*, 212.

81. Larry Sabato, *PAC Power: Inside the World of Political Action Committees* (New York: Norton, 1984), 135; R. Kenneth Godwin, *One Billion Dollars of Influence: The Direct Marketing of Politics* (Chatham, NJ: Chatham House, 1988), 131–36; Larry Sabato, *Paying for Elections: The Campaign Finance Thicket* (New York: Priority, 1989), 13.

82. Frank R. Baumgartner and Beth L. Leech, *Basic Interests: The Importance of Groups in Politics and in Political Science* (Princeton, NJ: Princeton University Press, 1998), 131.

83. Baumgartner et al., *Lobbying and Policy Change*, 79.

84. David C. Kimball et al., "Who Cares about the Lobbying Agenda?" *Interest Groups & Advocacy* 1, no. 1 (May 2012): 5–25.

85. On the importance of perceptions and framing in shaping foreign economic policy preferences, see Raymond Hicks, Helen V. Milner, and Dustin Tingley, "Trade Policy, Economic Interests, and Party Politics in a Developing Country: The Political Economy of CAFTA-DR," *International Studies Quarterly* 58, no. 1 (March 2014): 106–17.

86. George J. Borjas, *Immigration Economics* (Cambridge, MA: Harvard University Press, 2014).

87. Jens Hainmueller and Michael J. Hiscox, "Attitudes Toward Highly Skilled and Low-Skilled Immigration: Evidence from a Survey Experiment," *American Political Science Review* 104, no. 1 (February 2010): 61.

88. Gary Clyde Hufbauer, Theodore H. Moran, and Lindsay Oldenski, *Outward Foreign Direct Investment and US Exports, Jobs, and R&D: Implications for US Policy* (Washington, DC: Peterson Institute for International Economics, 2013).

89. Neil Malhotra, Yotam Margalit, and Cecilia Hyunjung Mo, "Economic Explanations for Opposition to Immigration: Distinguishing Between Prevalence and Conditional Impact," *American Journal of Political Science* 57, no. 2 (April 2013): 391–410; Brenton D. Peterson, Sonal S. Pandya, and David Leblang, "Doctors with Borders: Occupational Licensing as an Implicit Barrier to High Skill Migration," *Public Choice* 160, no. 1–2 (July 2014): 45–63.

90. Peterson, Pandya, and Leblang, "Doctors with Borders."

91. Baumgartner et al., *Lobbying and Policy Change*, 232.

92. Ken Kollman, *Outside Lobbying: Public Opinion and Interest Group Strategies* (Princeton, NJ: Princeton University Press, 1998).

93. Chase, "Moving Hollywood Abroad: Divided Labor Markets and the New Politics of Trade in Services," 660; Giovanni Facchini, Anna Maria Mayda, and Prachi Mishra, "Do Interest Groups Affect US Immigration Policy?" *Journal of International Economics* 85, no. 1 (September 2011): 114–28.

94. Laurie P. Milton, "An Identity Perspective on the Propensity of High-Tech Talent to Unionize," *Journal of Labor Research* 24, no. 1 (2003): 32; Robbert van het Kaar and Marianne Grünell, "Industrial Relations in the Information and Communications Technology Sector," *Eurofound*, August 27, 2001, www.eurofound.europa.eu/observatories/eurwork /comparative-information/industrial-relations-in-the-information-and-communications -technology-sector. See also Lazonick, *Sustainable Prosperity in the New Economy?*, 144–46.

95. On the challenges of mass mobilization, see Baumgartner et al., *Lobbying and Policy Change*, 156–57.

96. Janice Fine and Daniel J. Tichenor, "An Enduring Dilemma: Immigration and Organized Labor in Western Europe and the United States," in *The Oxford Handbook of the Politics of International Migration*, ed. Marc Rosenblum and Daniel J. Tichenor (Oxford: Oxford University Press, 2012), 532–72.

97. I am following Berry's definition here. See Jeffrey M. Berry, *The New Liberalism: The Rising Power of Citizen Groups* (Washington, DC: Brookings Institution, 1999), 2.

98. Berry, *The New Liberalism*; Baumgartner et al., *Lobbying and Policy Change*, 238. Although these studies are focused on U.S. politics, citizen groups are effective opponents of business in other contexts. See Andreas Dür, Patrick Bernhagen, and David Marshall, "Interest Group Success in the European Union: When (and Why) Does Business Lose?" *Comparative Political Studies* 48, no. 8 (2015): 951–83; Andreas Dür and Gemma Mateo, "Public Opinion and Interest Group Influence: How Citizen Groups Derailed the

Anti-counterfeiting Trade Agreement," *Journal of European Public Policy* 21, no. 8 (2014): 1199–1217.

99. Thomas T. Holyoke, "The Interest Group Effect on Citizen Contact with Congress," *Party Politics* 19, no. 6 (November 1, 2013): 937–40.

100. Kollman, *Outside Lobbying*, 58–100; Baumgartner et al., *Lobbying and Policy Change*, 127.

101. John R. Wright, *Interest Groups and Congress: Lobbying, Contributions and Influence* (Boston: Allyn and Bacon, 1996).

102. On the utility of combining different strands of liberal international relations theory, see Andrew Moravcsik, "Liberal Theories of International Law," in *Interdisciplinary Perspectives on International Law and International Relations: The State of the Art*, ed. Jeffrey L. Dunoff and Mark A. Pollack (Cambridge: Cambridge University Press, 2013), 85. On liberal theory more generally, see Andrew Moravcsik, "Taking Preferences Seriously: A Liberal Theory of International Politics," *International Organization* 51, no. 4 (1997): 513–53.

103. David A. Lake, "Open Economy Politics: A Critical Review," *The Review of International Organizations* 4, no. 3 (September 2009): 225.

104. Joseph M. Grieco, "Anarchy and the Limits of Cooperation: A Realist Critique of the Newest Liberal Institutionalism," in *Neorealism and Neoliberalism: The Contemporary Debate* (New York: Columbia University Press, 1993), 129; John J. Mearsheimer, *The Tragedy of Great Power Politics* (New York: W. W. Norton, 2001), 401–2.

105. Recent work has emphasized how a state's domestic policies on innovation can reflect national security concerns. See Weiss, *America Inc.?*; Taylor, *The Politics of Innovation*.

106. Dong Jung Kim, "Cutting Off Your Nose? A Reigning Power's Commercial Containment of a Military Challenger" (Ph.D. dissertation, University of Chicago, 2015); Dong Jung Kim, "Trading with the Enemy? The Futility of US Commercial Countermeasures Against the Chinese Challenge," *Pacific Review*, November 2, 2016, 1–20.

107. On the prevalence of these problems during the Cold War, see Mastanduno, *Economic Containment*.

108. Smith, *American Business and Political Power*, 9.

109. Baumgartner et al., *Lobbying and Policy Change*, 29–45.

110. David Nelson and Susan Webb Yackee, "Lobbying Coalitions and Government Policy Change: An Analysis of Federal Agency Rulemaking," *The Journal of Politics* 74, no. 2 (2012): 339–53.

111. Martin Gilens and Benjamin I. Page, "Testing Theories of American Politics: Elites, Interest Groups, and Average Citizens," *Perspectives on Politics* 12, no. 3 (2014): 575.

112. On the utility of process tracing for testing causal mechanisms, see Alexander George and Andrew Bennett, *Case Studies and Theory Development in the Social Sciences* (Cambridge, MA: MIT Press, 2005), 205–32.

113. If need be, the analysis can also take into account additional considerations, such as the relevant visa fees, whether admission is contingent on a labor market test, and whether spouses are allowed to work. See Lucie Cerna, "The EU Blue Card: Preferences, Policies, and Negotiations Between Member States," *Migration Studies* 2, no. 1 (March 1, 2014): 9–10.

3. THE SWINGING DOOR

1. National Science Foundation, *Science and Engineering Indicators 2016* (Arlington, VA: National Science Foundation, 2016), chapter 3, 101.

2. National Science Foundation, *Science and Engineering Indicators 2016*, chapter 3, 104.

3. Demetrios G. Papademetriou and Stephen Yale-Loehr, "Balancing Interests: Rethinking U.S. Selection of Skilled Immigrants" (Washington, DC: Carnegie Endowment for International Peace, 1996), 40–48.

4. Quoted in H. Rosemary Jeronimides, "The H-1B Visa Category: A Tug of War," *Georgetown Immigration Law Journal* 7 (1993): 369.

5. Papademetriou and Yale-Loehr, "Balancing Interests," 82.

6. On the origins of the H-1B program, see Jeronimides, "The H-1B Visa Category."

7. Bruce Morrison, testimony, *Hearing Before the Committee on the Judiciary of the United States House of Representatives Subcommittee on Immigration Policy and Enforcement*, 113th Cong. (March 5, 2013), 2.

8. Morrison initially supported a cap of 25,000 H-1B visas, but he subsequently increased this to 65,000 at the conference stage. Morrison recalls that the initial limit was based on inaccurate data regarding current usage from the Immigration and Naturalization Service and that he increased the limit in response to additional information from the executive branch indicating that current usage was between 30,000 and 40,000 visas per year. Bruce Morrison, former chair, House Judiciary Committee Subcommittee on Immigration, interview by author, February 27, 2016.

9. Bruce Morrison, interview. On the politics behind the Immigration Act of 1990 more generally, see Daniel J. Tichenor, *Dividing Lines: The Politics of Immigration Control in America* (Princeton, NJ: Princeton University Press, 2002), 267–74.

10. Daniel Costa, "Little-Known Temporary Visas for Foreign Tech Workers Depress Wages," *The Hill*, November 11, 2014, http://thehill.com/blogs/pundits-blog/technology/223607 -little-known-temporary-visas-for-foreign-tech-workers-depress.

11. Neil G. Ruiz, Jill H. Wilson, and Shyamali Choudhury, "The Search for Skills: Demand for H-1B Immigrant Workers in U.S. Metropolitan Areas" (Washington, DC: Brookings Institution, 2012), 4, 7, 14. These authors estimate that higher education, government, and non-profit research institutions accounted for about 10 percent of H-1B labor condition applications submitted between 2001 and 2010.

12. Ruth Ellen Wasem, *Immigration: Legislative Issues on Nonimmigrant Professional Specialty (H-1B) Workers* (Washington, DC: Congressional Research Service, 2004), 6.

13. U.S. Immigration and Naturalization Service, *Report on Characteristics of Specialty Occupation Workers (H-1B): Fiscal Year 2000* (Washington, DC: U.S. Immigration and Naturalization Service, 2002), 8.

14. U.S. General Accounting Office, *Immigration and the Labor Market: Nonimmigrant Alien Workers in the United States* (Washington, DC: U.S. General Accounting Office, 1992), 34, www.gao.gov/assets/160/151654.pdf.

15. U.S. Immigration and Naturalization Service, *Report on Characteristics*, 5.

16. U.S. Citizenship and Immigration Services, *Characteristics of H-1B Specialty Occupation Workers: Fiscal Year 2015* (Washington, DC: U.S. Department of Homeland Security, 2016), 8.

17. Liz Robbins, "New U.S. Rule Extends Stay for Some Foreign Graduates," *New York Times*, March 9, 2016, www.nytimes.com/2016/03/09/nyregion/new-us-rule-extends-stay-for-some -foreign-graduates.html.

18. Lynne Shotwell, executive director, Council for Global Immigration, interview by author, July 22, 2015.

19. Joan Sazabo, "Opening Doors for Immigrants," *Nation's Business*, August 1, 1989.

20. Daryl Buffenstein and Kevin Miner, Fragomen Worldwide, interview by author, February 12, 2016. Buffenstein is a former president of the American Immigration Lawyers Association (AILA) and testified before Congress regarding the Immigration Act of 1990. Miner is the vice-chair of the AILA's Liaison Committee with the U.S. Department of Labor. The quote is from Buffenstein.

21. Ronil Hira, "U.S. Immigration Regulations and India's Information Technology Industry," *Technological Forecasting and Social Change* 71, no. 8 (October 2004): 841.

22. The source for this and the previous sentence is Daryl Buffenstein and Kevin Miner, interview by author. The quotes are from Buffenstein.

23. Hira, "U.S. Immigration Regulations," 846.

24. American Immigration Lawyers Association, "Analysis of the American Competitiveness in the Twenty-First Century Act," *Shusterman.com*, accessed March 2, 2016, http://shusterman .com/h1b-analysisofac21.html.

25. Quoted in Cindy Rodriguez, "Foreign Workers Bill Approved," *Boston Globe*, October 4, 2000.

26. William Glanz, "High-Tech Lobbyist Counts Washington Successes," *Knight-Ridder Tribune Business News*, October 14, 2000.

27. Michael S. Teitelbaum, *Falling Behind? Boom, Bust, and the Global Race for Scientific Talent* (Princeton, NJ: Princeton University Press, 2014), 57–58.

28. Jason Zengerle, "Silicon Smoothies," *New Republic*, June 8, 1998.

29. Quoted in Glanz, "High-Tech Lobbyist Counts Washington Successes."

30. Zengerle, "Silicon Smoothies," 21.

31. John Simons, "Impasse on Bill to Boost Visas Persists Between Firms, U.S.," *Wall Street Journal*, August 6, 1998.

32. Zengerle, "Silicon Smoothies," 20.

33. Laurie P. Milton, "An Identity Perspective on the Propensity of High-Tech Talent to Unionize," *Journal of Labor Research* 24, no. 1 (2003): 32.

34. Quoted in Bill Pietrucha, "Labor Challenges High Tech Job Shortage Claims," *Newsbytes News Network*, March 19, 1998.

35. Paul Kostek, president, IEEE-USA, 1999, interview by author, June 29, 2015.

36. Ana Avendano, director of immigration, 2004 to 2009, and executive assistant to the president for immigration, 2009 to 2014, AFL-CIO, interview by author, February 21, 2015.

37. "FAIR Statement on the Pioneer Fund," *PR Newswire,* March 4, 1998; Devin Burghart and Leonard Zeskind, "Beyond FAIR: The Decline of the Established Anti-Immigration Organizations and the Rise of Tea Party Nativism" (Kansas City, MO: Institute for Research and Education on Human Rights, 2012), 4, www.irehr.org/issue-areas/tea-party-nationalism/beyond-fair-report.

38. Roy Beck, president, NumbersUSA, personal communication, June 18, 2015. See also "Lott Wants Agreement with Dems on H-1B Visa Measure," *Congress Daily,* September 14, 2000.

39. Carolyn Wong, *Lobbying for Inclusion: Rights Politics and the Making of Immigration Policy* (Stanford, CA: Stanford University Press, 2006), 133–45.

40. Quoted in Miranda Ewell, "Clinton Opposes Higher Visa Cap; Focus on 'Home-Grown' Talent, Commerce Chief Says," *San Jose Mercury News,* January 13, 1998.

41. Quoted in Ewell, "Clinton Opposes Higher Visa Cap."

42. U.S. General Accounting Office, *Assessment of the Department of Commerce's Report on Workforce Demand and Supply* (Washington, DC: U.S. General Accounting Office, 1998), 2, 6, http://gao.gov/assets/230/225415.pdf.

43. Pietrucha, "Labor Challenges High Tech Job Shortage Claims."

44. Simons, "Impasse on Bill to Boost Visas Persists Between Firms, U.S."

45. The safeguards limited the employer's ability to transfer an H-1B worker to another employer, required that employers certify that they had not and would not displace American workers, and obliged employers to take steps to recruit American workers who are equally or better qualified for the position for which an H-1B worker is sought. See Jessica F. Rosenbaum, "Exploiting Dreams: H-1B Visa Fraud, Its Effects, and Potential Solutions," *University of Pennsylvania Journal of Business Law* 13, no. 3 (2010): 802. Note that firms employing only "exempt H-1B workers" (those making at least $60,000 a year or having the equivalent of a Master's degree or higher) are not subject to these restrictions. See U.S. Department of Labor, "Fact Sheet 62: What Are 'Exempt' H-1B Nonimmigrants?" *Wage and Hour Division,* July 2008, 1, www.dol.gov/whd/regs/compliance/FactSheet62/whdfs62Q.pdf.

46. Tom Abate and Jon Swartz, "Eleventh-Hour Victory for Tech: Visa Increase, R&D Tax Measure in Budget Bill," *San Francisco Chronicle,* October 16, 1998, www.sfgate.com/business/article/11th-Hour-Victory-For-Tech-Visa-increase-R-D-2984825.php.

47. For a helpful summary of the legislative process and the content of the bill, see Jung S. Hahm, "American Competitiveness and Workforce Improvement Act of 1998: Balancing Economic and Labor Interests Under the New H-1B Visa Program," *Cornell Law Review* 85 (1999): 1683–88.

48. Mark Leibovich, "High Tech Is King of the Hill," *Washington Post,* October 16, 1998.

49. Quoted in Leibovich, "High Tech Is King of the Hill."

50. Quoted in Abate and Swartz, "Eleventh-Hour Victory for Tech: Visa Increase, R&D Tax Measure in Budget Bill."

51. William Branigin, "Visa Deal for Computer Programmers Angers Labor Groups," *Washington Post,* September 27, 1998.

52. Quoted in Branigin, "Visa Deal for Computer Programmers Angers Labor Groups."
53. Quoted in Branigin, "Visa Deal for Computer Programmers Angers Labor Groups."
54. Harris Miller, president, Information Technology Association of America, 1995 to 2005, interview by author, July 1, 2015.
55. Teitelbaum, *Falling Behind?*, 113–14.
56. American Federation of Labor and Congress of Industrial Organizations (AFL-CIO), "Immigration," *AFL-CIO*, February 16, 2000, www.aflcio.org/About/Exec-Council/EC -Statements/Immigration.
57. "House Dems, AFL-CIO Discuss H-1B Visas," *Congress Daily*, March 14, 2000.
58. Jube Shiver, "Alliance Fights Boost in Visas for Tech Workers," *Los Angeles Times*, August 5, 2000, http://articles.latimes.com/2000/aug/05/business/fi-64994.
59. Terry Costlow, "Senate Set to Vote This Week on Visa-Cap Bill; High Noon Approaches for H-1B Friends, Foes," *Electronic Engineering Times*, October 2, 2000.
60. Paul Kostek, interview.
61. Robert Pear, "Clinton Asks Congress to Raise the Limit on Visas for Skilled Workers," *New York Times*, May 12, 2000.
62. Susan Bibler Coutin, *Nations of Emigrants: Shifting Boundaries of Citizenship in El Salvador and the United States* (Ithaca, NY: Cornell University Press, 2007), 184–87.
63. Bibler Coutin, *Nations of Emigrants*, 188.
64. American Competitiveness in the Twenty-First Century Act of 2000, Pub. L. No. 106-313, 114 Stat. 1251 (2000), www.gpo.gov/fdsys/pkg/BILLS-106s2045enr/pdf/BILLS-106s2045enr .pdf. See also Carolyn Lochhead, "Bill to Boost Tech Visas Sails Through Congress," *San Francisco Chronicle*, October 4, 2000, www.sfgate.com/news/article/Bill-to-Boost-Tech -Visas-Sails-Through-Congress-2735682.php.
65. Quoted in Lochhead, "Bill to Boost Tech Visas Sails Through Congress." See also Norman Matloff, "On the Need for Reform of the H-1B Non-Immigrant Work Visa in Computer- Related Occupations," *University of Michigan Journal of Law Reform* 36, no. 4 (Fall 2003): 815–914.
66. Quoted in "Analysis: Look at the Controversy Over H-1B Visas," *National Public Radio: Talk of the Nation*, September 26, 2000.
67. Quoted in Costlow, "Senate Set to Vote This Week on Visa-Cap Bill."
68. Quoted in "Analysis: Look at the Controversy Over H-1B Visas."
69. Marjorie Valbrun and Scott Thurm, "Foreign Workers Will Soon Get Fewer U.S. Visas," *Wall Street Journal*, October 1, 2003.
70. Jena Heath, "Congressman Switches Focus to High-Tech: Republican, New to Area, Softens Immigration Stance, Trumpets Tech," *Austin American-Statesman*, July 7, 2002.
71. Sandra Boyd, chair, Compete America, 2004 to 2006, interview by author, June 29, 2015.
72. Sara Schaefer Munoz, "Firms Push to Expand Visa Program," *Wall Street Journal*, March 11, 2004; Teitelbaum, *Falling Behind?*, 109.
73. Sandra Boyd, interview.
74. Danielle Belopotosky, "Lobbyists Push Congress for Action On H-1B Visas," *Technology Daily*, November 16, 2004.

75. Elizabeth Olson, "Congress Raises Limit on Skilled-Work Visas," *International Herald Tribune*, November 24, 2004.

76. Danielle Belopotosky, "Efforts to Change Visa Law for Skilled Workers Gains Steam," *Technology Daily*, October 1, 2004; Chloe Albanesius, "Outsourcing Controversy Influences Debate On H-1B Visas," *Technology Daily*, May 6, 2004.

77. Michael Posner, "Groups Jockey for Position On Possible Boost In H-1B Visas," *Congress Daily*, November 17, 2004.

78. Personal communication from Roy Beck, president, NumbersUSA, June 18, 2015.

79. Roy Beck, president, NumbersUSA, interview by author, June 13, 2015.

80. For a summary, see Richard Rulon, "Competing for Foreign Talent," *Legal Intelligencer*, December 15, 2004.

81. Glen Kessler and Kevin Sullivan, "Powell Cautious About Immigration Changes," *Washington Post*, November 10, 2004.

82. Michael Fletcher, "Bush Immigration Plan Meets GOP Opposition: Lawmakers Resist Temporary-Worker Proposal," *Washington Post*, January 2, 2005.

83. Michael Fletcher, "Bush Immigration Plan Meets GOP Opposition."

84. "Senator Kerry Delivers Democratic Hispanic Radio Address," *U.S. Fed News*, April 1, 2006.

85. "Compete America," *Technology Daily PM*, October 4, 2005.

86. "Compete America"; Edward Alden, "Emigrants to US Face Long Wait for Green Card," *Financial Times*, October 12, 2005.

87. Patrick O'Connor, "Anti-Immigration Groups Up Against Unusual Coalition," *The Hill*, February 28, 2006; Kate Phillips, "Business Lobbyists Call for Action on Immigration," *New York Times*, April 15, 2006, www.nytimes.com/2006/04/15/us/15lobby.html.

88. Nina Bernstein, "In the Streets, Suddenly, an Immigrant Groundswell," *New York Times*, March 27, 2006; Gerardo Lissardy, "Leading Hispanic Group Joins Immigrant-Rights Coalition," *EFE News Service*, May 12, 2006.

89. Jennifer Ludden, "Strange Bedfellows Join Forces for Immigration Reform," *National Public Radio: All Things Considered*, January 19, 2006.

90. Roy Beck, personal communication, June 18, 2015.

91. "NumbersUSA Activists Squash Amnesty in Senate: Senate Rejects Cloture on Amnesty Bill 46–53," *PR Newswire*, June 28, 2007.

92. Molly Ball, "The Little Group Behind the Big Fight to Stop Immigration Reform," *The Atlantic*, August 1, 2013, www.theatlantic.com/politics/archive/2013/08/the-little -group-behind-the-big-fight-to-stop-immigration-reform/278252/.

93. Quoted in Brian Mitchell, "Frist's Border Control-Only Bill Spurs Broad Immigration Deals," *Investor's Business Daily*, March 20, 2006.

94. Quoted in Mitchell, "Frist's Border Control-Only Bill Spurs Broad Immigration Deals."

95. Comprehensive Immigration Reform Act of 2006, S.2611, 109th Cong. (2006), www .congress.gov/bill/109th-congress/senate-bill/2611. See also Wasem, *Immigration*, 24–25.

96. Quoted in Rachel L. Swarns, "Senate, in Bipartisan Act, Passes an Immigration Bill," *New York Times*, May 26, 2006, www.nytimes.com/2006/05/26/washington/26immig.html.

97. Suzanne Gamboa, "Senate Vote Sidetracks Immigration Compromise," *Associated Press*, April 7, 2006.

98. Jonathan Weisman and Jim VandeHei, "Immigration Bill Lobbying Focuses on House Leaders: With Senate in Hand, Bush May Face a Skeptical GOP Base," *Washington Post*, May 1, 2006.

99. Roy Beck, interview, April 3, 2015.

100. "Address by House Speaker J. Dennis Hastert: Reflections on Role of Speaker in Modern Day House of Representatives," *U.S. Newswire*, November 12, 2003.

101. Anne C. Mukem, "Firebrand Tancredo Puts Policy Over Party Line," *Denver Post*, November 27, 2005; "Rep. Tancredo Slams Senate's Compromise on Amnesty," *U.S. Fed News*, May 11, 2006.

102. Except as noted, the following sentences are based on Roy Beck, interview, June 13, 2015.

103. Southern Poverty Law Center, "John Tanton Is the Mastermind Behind the Organized Anti-Immigration Movement," *Intelligence Report*, no. 106 (Summer 2002), www .splcenter.org/get-informed/intelligence-report/browse-all-issues/2002/summer/the -puppeteer?page=0,3.

104. Roy Beck, interview, June 13, 2015.

105. Jim VandeHei and Zachary A. Goldfarb, "Immigration Deal at Risk as House GOP Looks to Voters," *Washington Post*, May 28, 2006.

106. Charles Babington, "Immigration Bill Expected to Pass Senate This Week: Hastert May Block Version That Divides House GOP," *Washington Post*, May 23, 2006.

107. Quoted in Babington, "Immigration Bill Expected to Pass Senate This Week."

108. Mike Madden, "Millions Spent Lobbying on Immigration in Last Congress," *Gannett News Service*, January 26, 2007.

109. Quoted in Reinhardt Krause, "Tech Firms Pushing for More H-1B Visas for Skilled Workers," *Investor's Business Daily*, April 11, 2007.

110. Eunice Moscoso, "Once Gung-Ho, Businesses See Flaws in Immigration Bill: Tech Sector Particularly Disturbed by Potential Changes in Visa Program," *Atlanta Journal-Constitution*, June 3, 2007.

111. Carolyn Lochhead, "Visa Plan Angers Silicon Valley," *SFGate*, June 7, 2007, www.sfgate .com/politics/article/VISA-PLAN-ANGERS-SILICON-VALLEY-Immigration -2588829.php.

112. Ana Avendano, interview, May 25, 2017.

113. Amanda Paulson, Faye Bowers, and Daniel Wood, "To Immigrants, US Reform Bill Is Unrealistic," *Christian Science Monitor*, May 21, 2007.

114. Quoted in Ryan Lizza, "Getting to Maybe," *New Yorker*, June 24, 2013, www.newyorker .com/magazine/2013/06/24/getting-to-maybe.

115. Shailagh Murray, "Careful Strategy Is Used to Derail Immigration Bill," *Washington Post*, June 8, 2007.

116. John Stanton and Jennifer Yachnin, "Reid Plots to Block Conservatives," *Roll Call*, June 18, 2007.

117. Robert Pear, "U.S. High-Tech Firms Stymied on Immigration for Skilled Workers," *New York Times*, June 25, 2007, www.nytimes.com/2007/06/25/technology/25iht-visas .4.6326165.html.

118. Daryl Buffenstein and Kevin Miner, interview.

119. Kathy Kiely, "Immigration Overhaul Crumbles in Senate Vote," *USA Today*, June 29, 2007.

120. Jonathan Weisman, "Immigration Bill Dies in Senate," *Washington Post*, June 29, 2007, www.washingtonpost.com/wp-dyn/content/article/2007/06/28/AR2007062800963 .html.

121. Congressional aide, interview by author, May 24, 2017.

122. Quoted in Kiely, "Immigration Overhaul Crumbles in Senate Vote."

123. Dana Milbank, "Jabs and All, the Ides of March Arrives Late," *Washington Post*, June 29, 2007.

124. Roy Beck, interview by author, April 3, 2015. See also Bruce Schreiner, "National Group Takes Aim at McConnell on Immigration," *Associated Press*, June 27, 2007.

125. Quoted in Nicole Gaouette, "Immigration Bill Ignites a Grass-Roots Fire on the Right," *Los Angeles Times*, June 24, 2007, http://articles.latimes.com/2007/jun/24/nation /na-immig24.

126. Stuart Rothenberg, "Heeee's Back: The Fall and Rise of Sen. Trent Lott," *Roll Call*, May 22, 2006.

127. Milbank, "Jabs and All, the Ides of March Arrives Late."

128. Mitch McConnell (KY), "Immigration," *Congressional Record*, 110th Cong., 153, no. 106 (June 28, 2007): S8674.

129. Roy Beck, interview, April 3, 2015.

130. Quoted in Heather Greenfield, "Techies 'Disappointed' by Immigration Bill's Demise," *Technology Daily PM*, June 28, 2007.

131. Quoted in Robert Pear, "Little-Known Group Claims a Win on Immigration," *New York Times*, July 15, 2007.

132. Quoted in Gaouette, "Immigration Bill Ignites a Grass-Roots Fire on the Right."

133. Quoted in Molly Ball, "Immigration Reformers Are Winning August," *The Atlantic*, August 21, 2013, www.theatlantic.com/politics/archive/2013/08/immigration-reformers-are -winning-august/278873/.

134. Nicholas Thompson, "Obama vs. McCain: The Wired.com Scorecard," *Wired*, October 12, 2008, www.wired.com/2008/10/obama-v-mccain/.

135. Quoted in Neil Munro, "IT Industry, Hispanics Team Up On Immigration," *National Journal*, April 9, 2010.

136. Quoted in Munro, "IT Industry, Hispanics Team Up On Immigration."

137. Quoted in Munro, "IT Industry, Hispanics Team Up On Immigration."

138. Terry Hartle, senior vice-president, and Steven Bloom, director for federal relations, American Council on Education, interview by author, September 8, 2015. The quote is from Hartle.

139. Munro, "IT Industry, Hispanics Team Up On Immigration."

140. National Council of La Raza, *2011 Annual Report* (Washington, DC: National Council of La Raza, 2011), http://publications.nclr.org/handle/123456789/2; "Hispanic Business Magazine Includes 12 NCLR Affiliates Among Its Top 25 Nonprofits," *Targeted News Service*, June 9, 2011.

141. Mark Hugo Lopez, "The Hispanic Vote in the 2008 Election" (Washington, DC: Pew Research Center, November 5, 2008), www.pewhispanic.org/2008/11/05/the-hispanic-vote-in-the-2008-election/; Eric Schurenberg, "Why the Next Steve Jobs Could Be an Indian," *Mint*, October 28, 2011.

142. Munro, "IT Industry, Hispanics Team Up On Immigration."

143. "Sen. Chuck Grassley to Place Hold on Employment-Based Visa Bill," *NumbersUSA*, November 30, 2011, www.numbersusa.com/content/news/november-30-2011/sen-chuck-grassley-place-hold-employment-based-visa-bill.html.

144. David Bier, "High-Skilled Immigration Restrictions Are Economically Senseless," *Forbes*, July 22, 2012, www.forbes.com/sites/realspin/2012/07/22/high-skilled-immigration-restrictions-are-economically-senseless/; David Bier, "Why Does the Government Care Where Immigrant Workers Were Born?" *Cato Institute*, January 18, 2017, www.cato.org/blog/why-does-government-care-where-immigrant-workers-were-born.

145. Former congressional aide, interview by author, June 9, 2017.

146. Laura Meckler, "Visas Could Aid Graduates," *Wall Street Journal*, October 22, 2011.

147. Bruce Morrison, interview. Morrison was representing the IEEE-USA at the time.

148. Roy Beck, interview, April 3, 2015.

149. Chris Frates, "Why the Schumer–Smith Immigration Negotiations Broke Down," *The Atlantic*, September 20, 2012, www.theatlantic.com/politics/archive/2012/09/why-the-schumer-smith-immigration-negotiations-broke-down/428835/.

150. Suzy Khimm, "Why a Rare Bipartisan Consensus on Immigration Totally Fell Apart," *Washington Post*, September 21, 2012, www.washingtonpost.com/blogs/ezra-klein/wp/2012/09/21/why-a-rare-bipartisan-consensus-on-immigration-totally-fell-apart/.

151. Frates, "Why the Schumer–Smith Immigration Negotiations Broke Down."

152. Barack Obama, "Statement of Administration Policy: H.R. 6429—STEM Jobs Act of 2012," *American Presidency Project*, November 28, 2012, www.presidency.ucsb.edu/ws/?pid=102707.

153. Gregory Wallace and Deirdre Walsh, "House Passes Immigration Bill to Keep Science and Technology Students in U.S.," *CNN Wire*, November 30, 2012.

154. Quoted in Brian Fung, "Democrats' Dilemma on High-Skilled Immigration Reform," *National Journal*, March 6, 2013.

155. Ana Avendano, interview, February 21, 2015. See also Richard Trumka and Thomas J. Donohue, "Joint Statement of Shared Principles by U.S. Chamber of Commerce President and CEO Thomas J. Donohue and AFL-CIO President Richard Trumka," *AFL-CIO*, February 21, 2013, www.aflcio.org/Press-Room/Press-Releases/Joint-Statement-of-Shared-Principles-by-U.S.-Chamber-of-Commerce-President-and-CEO-Thomas-J.-Donohue-AFL-CIO-President-Richard-Trumka.

156. Lizza, "Getting to Maybe"; Jason Horowitz, "Marco Rubio Pushed for Immigration Reform with Conservative Media," *New York Times*, February 27, 2016, www.nytimes.com/2016/02/28/us/politics/marco-rubio-pushed-for-immigration-reform-with-conservative-media.html.

157. Center for Responsive Politics, "Lobbying Spending Database: Microsoft," *OpenSecrets. org*, 2015, www.opensecrets.org/lobby/clientsum.php?id=D000000115&year=2013.

158. Eric Lipton and Somini Sengupta, "Latest Product of Tech Firms: Immigrant Bill," *New York Times*, May 5, 2013.

159. Mark Zuckerberg, "Mark Zuckerberg: Immigrants Are the Key to a Knowledge Economy," *Washington Post*, April 10, 2013, www.washingtonpost.com/opinions/mark-zuckerberg-immigrants-are-the-key-to-a-knowledge-economy/2013/04/10/aba05554-a20b-11e2-82bc-511538ae90a4_story.html.

160. Border Security, Economic Opportunity, and Immigration Modernization Act of 2013, S. 744, 113th Cong. (2013), www.govtrack.us/congress/bills/113/s744.

161. Border Security, Economic Opportunity, and Immigration Modernization Act of 2013.

162. The original bill required companies using H-1B visas to first advertise job openings on a government website for thirty days, to extend offers to Americans before hiring a visa holder, and to show they had not laid off an American worker within three months of hiring a visa holder from overseas. The bill also gave the Department of Labor the power to audit and challenge companies' hiring decisions for up to two years after they were made. See Fredreka Schouten, "Tech Firms Would Skirt Hiring Restrictions Under Deal," *USA Today*, May 21, 2013; Lipton and Sengupta, "Latest Product of Tech Firms."

163. Peter Walsten, Jia Lynn Yang, and Craig Timberg, "Facebook Flexes Political Muscle with Carve-Out in Immigration Bill," *Washington Post*, April 16, 2013.

164. Ana Avendano, interview, February 21, 2015.

165. Ana Avendano, personal communication, May 30, 2017.

166. Ronil Hira, testimony, *Hearing Before the Senate Judiciary Committee on the Border Security, Economic Opportunity, and Immigration Modernization Act, S.744*, 113th Cong. (April 22, 2013), 10; Paul Roy, "Impact of U.S. Senate Bill on Outsourcing," *Mondaq Business Briefing*, July 31, 2013.

167. Dhanya Ann Thoppil and Sean McLain, "Q&A: 'Parts of U.S. Visa Bill Discriminatory,'" *Wall Street Journal*, April 26, 2013. Indian firms had complained of discrimination in 2010 after the United States raised H-1B and L-1 visa fees for companies with at least fifty employees, at least 50 percent of whom were on either H-1B or L-1 visas.

168. Quoted in Lipton and Sengupta, "Latest Product of Tech Firms."

169. "Technology Leaders Urge U.S. Senate to Approve Comprehensive Immigration Reform Legislation," *Information Technology Industry Council*, June 20, 2013, www.itic.org/news-events/news-releases/technology-leaders-urge-u-s-senate-to-approve-comprehensive-immigration-reform-legislation.

170. American Council on Education, "Higher Education Associations Strongly Endorse Senate Immigration Reform Bill" (letter to the U.S. Senate), June 26, 2013, www.aau.edu/sites/default/files/AAU%20Files/Key%20Issues/Budget%20%26%20Appropriations/FY17/Endorsement-Letter-S-744_6-26-13.pdf.

171. Ball, "Immigration Reformers Are Winning August."

172. Zachary A. Goldfarb and Karen Tumulty, "IRS Admits Targeting Conservatives for Tax Scrutiny in 2012 Election," *Washington Post*, May 10, 2013, www.washingtonpost.com /business/economy/irs-admits-targeting-conservatives-for-tax-scrutiny-in-2012-election /2013/05/10/3b6a0ada-b987-11e2-92f3-f291801936b8_story.html.

173. Devin Burghart, *Special Report: The Status of the Tea Party Movement—Part Two* (Kansas, MO: Institute for Research and Education on Human Rights, 2014), www.irehr .org/2014/01/21/status-of-tea-party-by-the-numbers/.

174. Rebecca Kaplan, "How the Tea Party Came Around on Immigration," *National Journal*, March 21, 2013.

175. Jill Lawrence, "The Myth of Marco Rubio's Immigration Problem," *National Journal*, July 15, 2013.

176. Lawrence, "The Myth of Marco Rubio's Immigration Problem."

177. Linda Feldmann, "Is the Tea Party Running out of Steam?" *Christian Science Monitor*, April 12, 2014.

178. Tea Party Patriots, "Senate Must Admit Full Costs of Immigration Bill Before Passing Another 'Train Wreck,'" *Tea Party Patriots*, May 6, 2013, www.teapartypatriots.org/all-issues /news/senate-must-admit-full-costs-of-immigration-bill-before-passing-another-train-wreck/.

179. Katherine Rosario, "Five Simple Signs the Senate Immigration Bill Is Bad News," *Heritage Action*, April 17, 2013, http://heritageaction.com/2013/04/5-simple-signs-the -senate-immigration-bill-is-bad-news/; David Nakamura, "Conservatives Split on Immigration Bill's Price Tag," *Washington Post*, May 7, 2013.

180. Daniel Horowitz, "Gang Immigration Bill (S.744) Is Comprehensively Flawed," *Madison Project*, May 1, 2013, http://madisonproject.com/2013/05/gang-immigration-bill-s-744-is -comprehensively-flawed/.

181. Quoted in Lawrence, "The Myth of Marco Rubio's Immigration Problem."

182. Burghart and Zeskind, "Beyond FAIR," 23.

183. Roy Beck, interview, April 3, 2015.

184. Devin Burghart, "Mapping the Tea Party Caucus in the 112th Congress," *Institute for Research and Education on Human Rights*, March 17, 2011, https://irehr.org/issue-areas/tea -party-nationalism/tea-party-news-and-analysis/item/355-mapping-the-tea-party -caucus-in-the-112th-congress.

185. Laura Meckler, "Immigration-Bill Pressure Backfires: Overhaul Backers Target Majority Whip, but Tactic Provokes Response from Opponents," *Wall Street Journal*, December 25, 2013, www.wsj.com/articles/SB10001424052702304244904579276403694719232.

186. Quoted in Meckler, "Immigration-Bill Pressure Backfires."

187. Meckler, "Immigration-Bill Pressure Backfires."

188. Michael D. Shear and Ashley Parker, "Boehner Is Said to Back Change on Immigration," *New York Times*, January 1, 2014, www.nytimes.com/2014/01/02/us/politics/boehner-is -said-to-back-change-on-immigration.html.

189. Deirdre Shesgreen, "Immigration Reform Critics Blast Boehner's Remarks," *Gannett News Service*, April 25, 2014.

190. Roy Beck, interview, April 3, 2015.

191. Janet Hook, "Tea Party Faces Test of Its Clout in Primaries," *Wall Street Journal*, February 25, 2014.

192. Steven Rosenfeld, "The GOP's Vicious Internal War: Republican Establishment Trying to Exile Tea Partiers and Extremists," *AlterNet*, February 12, 2014, www.alternet .org/tea-party-and-right/gops-vicious-internal-war-republican-establishment-trying -exile-tea-partiers-and.

193. Craig Lindwarm, Director for Congressional and Governmental Affairs, Association for Public and Land Grant Universities, interview by author, September 11, 2015.

194. Scott Corley, executive director, Compete America, interview by author, May 18, 2017.

195. Jonathan Weisman, "Boehner Doubts Immigration Bill Will Pass in 2014," *New York Times*, February 6, 2014, www.nytimes.com/2014/02/07/us/politics/boehner-doubts -immigration-overhaul-will-pass-this-year.html.

196. Shear and Parker, "Boehner Is Said to Back Change on Immigration."

197. Quoted in Laura Meckler, "House Immigration Bills Are Still in the Mix," *Wall Street Journal*, April 18, 2014, www.wsj.com/articles/SB100014240527023046263045795080918395 46088.

198. Shesgreen, "Immigration Reform Critics Blast Boehner's Remarks."

199. Gail Russell Chaddock, "Eric Cantor Upset Stuns GOP, Revives Tea Party," *Christian Science Monitor*, June 11, 2014, www.csmonitor.com/USA/Elections/2014/0611/Eric-Cantor -upset-stuns-GOP-revives-tea-party.

200. The APLU, for example, proposed giving preference to graduates of U.S. universities in the allocation of H-1B visas. Craig Lindwarm, interview.

201. Obama administration White House official, interview by author, April 14, 2017.

202. Office of the Press Secretary, "Fact Sheet: Immigration Accountability Executive Action," *White House*, November 20, 2014, https://obamawhitehouse.archives.gov/the-press-office /2014/11/20/fact-sheet-immigration-accountability-executive-action.

203. Felicia Escobar Carrillo, special assistant to the president for immigration policy, 2014 to 2017, interview by author, May 18, 2017.

204. Quoted in Jessica Meyers, "Tech Companies See Few Big Gains in Obama's Executive Action," *Boston Globe*, November 24, 2014, www.bostonglobe.com/news/nation/2014/11 /24/tech-companies-see-few-big-gains-obama-executive-action/dauDJujkOhe1qx5 ZQTScoM/story.html.

205. Chris Currie, "U.S. Public Overwhelmingly Opposed to H-1B Visa Expansion," *IEEE-USA*, September 16, 1998, www.ieeeusa.org/communications/releases/_private/1998/pr091698 .html.

206. Jens Hainmueller and Michael J. Hiscox, "Attitudes Toward Highly Skilled and Low-Skilled Immigration: Evidence from a Survey Experiment," *American Political Science Review* 104, no. 1 (February 2010): 7.

207. Andrew Dugan, "Passing New Immigration Laws Is Important to Americans," *Gallup. com*, July 11, 2013, www.gallup.com/poll/163475/passing-new-immigration-laws-important -americans.aspx.

4. THE OPEN DOOR

1. Organisation for Economic Co-operation and Development (OECD), *Education at a Glance 2014: OECD Indicators* (Paris: OECD, 2014), 344.
2. OECD, *Education at a Glance 2014*, 344.
3. Institute for International Education, "Top Twenty-Five Places of Origin of International Students, 2014/15 and 2015/16," *Open Doors Report on International Educational Exchange* (New York: Institute for International Education, 2016), www.iie.org/opendoors.
4. Todd M. Davis, *Open Doors 2000: Report on International Education Exchange* (New York: Institute for International Education, 2000), 8; Hey-Kyung Koh Chin, *Open Doors 2002: Report on International Educational Exchange* (New York: Institute for International Education, 2002), 8.
5. Patricia Chow and Rajika Bhandari, *Open Doors 2010: Report on International Educational Exchange* (New York: Institute for International Education, 2010), 5.
6. Except as noted, this chapter follows the National Science Foundation's practice and considers S&E to include the physical sciences (including mathematics and computer sciences), natural sciences, social sciences, and engineering.
7. Chad C. Haddal, *Foreign Students in the United States: Policies and Legislation* (Washington, DC: Congressional Research Service, 2008).
8. U.S. Department of State, "Nonimmigrant Visa Issuances by Visa Class and by Nationality," *Travel.State.Gov*, accessed September 9, 2016, https://travel.state.gov/content/visas/en /law-and-policy/statistics/non-immigrant-visas.html.
9. Paul Stephens, "International Students: Separate but Profitable," *Washington Monthly*, October 2013, www.washingtonmonthly.com/magazine/september_october_2013/features /international_students_separato46454.php?page=all.
10. Institute for International Education, *Open Doors Data: Fast Facts* (New York: Institute for International Education, 2014), www.iie.org/Research-and-Publications/Open-Doors /Data/Fast-Facts; Tamar Lewin, "Foreign Students Bring Cash, and Changes: U.S. Colleges Welcome Funds, but Some In-State Applicants Feel Left Out," *International Herald Tribune*, February 6, 2012.
11. National Science Foundation, *Science and Engineering Indicators 2016* (Arlington, VA: National Science Foundation, 2016), chapter 2, 71.
12. National Science Foundation, *Science and Engineering Indicators 2016*, chapter 2, appendix table 27.
13. National Science Foundation, *Science and Engineering Indicators 2016*, chapter 2, appendix table 27.
14. Gnanaraj Chellaraj, Keith E. Maskus, and Aaditya Mattoo, "The Contribution of International Graduate Students to US Innovation," *Review of International Economics* 16, no. 3 (August 2008): 444–62.
15. Eric T. Stuen, Ahmed Mushfiq Mobarak, and Keith E. Maskus, "Skilled Immigration and Innovation: Evidence from Enrolment Fluctuations in US Doctoral Programmes," *The Economic Journal* 122, no. 565 (2012): 1143–76.

16. National Science Foundation, "Table 53: Doctorate Recipients with Temporary Visas Intending to Stay in the United States After Doctorate Receipt, by Country of Citizenship: 2007–13," *Science and Engineering Doctorates*, December 2014, www.nsf.gov/statistics /sed/2013/data-tables.cfm.

17. Dick Startz, "Sealing the Border Could Block one of America's Crucial Exports: Education," *The Brookings Institution*, January 31, 2017, https://www.brookings.edu/blog /brown-center-chalkboard/2017/01/31/sealing-the-border-could-block-one-of-americas -crucial-exports-education/.

18. Harris Miller, president, Information Technology Association of America, 1995 to 2005, interview by author, July 1, 2015.

19. Terry Hartle, senior vice-president, and Steven Bloom, director for federal relations, American Council on Education, interview by author, Washington, DC, September 8, 2015. The point was made by Hartle.

20. Philip G. Altbach and Patti McGill Peterson, "Internationalize American Higher Education? Not Exactly," *Change* 30, no. 4 (1998): 39.

21. Altbach and McGill Peterson, "Internationalize American Higher Education?," 38.

22. Nicholas Confessore, "Borderline Insanity," *Washington Monthly*, May 2002, www .washingtonmonthly.com/features/2001/0205.confessore.html.

23. Michael S. Teitelbaum, *Falling Behind? Boom, Bust, and the Global Race for Scientific Talent* (Princeton, NJ: Princeton University Press, 2014), 112.

24. Robert Farley, "9/11 Hijackers and Student Visas," *Factcheck.org*, May 10, 2013, www .factcheck.org/2013/05/911-hijackers-and-student-visas/.

25. The applicants were notified of their successful applications prior to the September 11 attacks. The flight school they attended in Florida received the formal approval letters in March 2002, however, underscoring the dysfunctionality of the U.S. immigration system. See Mark Potter and Rich Philips, "Six Months after Sept. 11, Hijackers' Visa Approval Letters Received," *CNN*, March 13, 2002, http://edition.cnn.com/2002/US/03/12/inv .flight.school.visas/.

26. Thomas B. Edsall, "Attacks Alter Politics, Shift Focus of Immigration Debate," *Washington Post*, October 15, 2001.

27. Terry Hartle and Steven Bloom, interview.

28. "Senator Feinstein Urges Major Changes in U.S. Student Visa Program," *Advocacy and Public Policymaking*, September 27, 2001, http://lobby.la.psu.edu/_107th/119_Student_Visas _Security/Congressional_Statements/Senate/S_Feinstein_09272001.htm.

29. Terry Hartle and Steven Bloom, interview. The quote is from Hartle.

30. Cindy Rodriguez, "Proposed Visa Ban Dropped," *Boston Globe*, November 23, 2001.

31. David Ward, "Letter to the Senate Judiciary Committee Regarding Feinstein Proposal on Student Visas," *American Association of Collegiate Registrars and Admissions Officers*, October 2, 2001, www.aacrao.org/advocacy/issues-advocacy/sevis.

32. Mark Sherman, "Feinstein Says Moratorium on Student Visas May Not Be Necessary," *Associated Press*, October 6, 2001.

33. Sherman, "Feinstein Says Moratorium on Student Visas May Not Be Necessary."

34. Rodriguez, "Proposed Visa Ban Dropped."

35. Sherman, "Feinstein Says Moratorium on Student Visas May Not Be Necessary."

36. Terry Hartle and Steven Bloom, interview. See also David Ward, "The Role of Technology in Preventing the Entry of Terrorists Into the United States," testimony, *Hearing Before the Subcommittee on Technology, Terrorism, and Government Information of the Senate Judiciary Committee*, 107th Cong. (October 12, 2001), 69.

37. Ward, "Letter to the Senate Judiciary Committee."

38. Cindy Rodriguez, "Congress Drops Plan to Bar Foreign Students," *Knight-Ridder Tribune Business News*, November 23, 2001.

39. Tucker Carlson and Bill Press, "Debating Immigration Policy," *Crossfire* (CNN, October 24, 2001).

40. Daniel Stein, president, Federation for American Immigration Reform, interview by author, July 2, 2015.

41. Kevin Drew, "Terror Probe Reaches Nation's Campuses," *CNN*, October 25, 2001, http://edition.cnn.com/2001/LAW/10/24/inv.international.students/index.html.

42. Center for Immigration Studies, "Are Foreign Students Good for America?" (panel discussion transcript, Rayburn Building, House of Representatives, Washington, DC, June 25, 2002), www.cis.org/sites/cis.org/files/articles/2002/foreignstudents.html.

43. Terry Hartle and Steven Bloom, interview.

44. Stephen Yale-Loehr, Demetrios G. Papademetriou, and Betsy Cooper, *Secure Borders, Open Doors: Visa Procedures in the Post–September 11 Era* (Washington, DC: Migration Policy Institute, 2005), 176.

45. Yale-Loehr, Papademetriou, and Cooper, *Secure Borders, Open Doors*, 178.

46. Mark Clayton, "Academia Becomes Target for New Security Laws," *Christian Science Monitor*, September 24, 2002.

47. David Ward, "Dealing with Foreign Students and Scholars in an Age of Terrorism: Visa Backlogs and Tracking Systems," testimony, *Hearing Before the U.S. House of Representatives Committee on Science*, 108th Cong. (March 26, 2003), 23.

48. Caryle Murphy and Nurith C. Aizenman, "Foreign Students Navigate Labyrinth of New Laws: Slip-Ups Overlooked Before 9/11 Now Grounds for Deportation," *Washington Post*, June 9, 2003.

49. Yale-Loehr, Papademetriou, and Cooper, *Secure Borders, Open Doors*, 178.

50. On the last two sentences, see Yale-Loehr, Papademetriou, and Cooper, *Secure Borders, Open Doors*, 177–78.

51. Lia Steakley, Debra K. Rubin, and Peter Reina, "After 9/11, Overseas Students Find Foreigners Need Not Apply: Visa Application Hurdles Start to Ease but Long-Term Impacts Loom," *Engineering News-Record*, December 6, 2004.

52. Andrew K. Collier, "Yale Chief Hits at US Student Visa Delays," *South China Morning Post*, November 14, 2003.

53. Murray Hiebert, "United States: The Cost of Security," *Far Eastern Economic Review*, November 28, 2002.

54. James Hattori, "Intel Plans for Future," *CNN*, September 14, 2002.

55. Martin Jischke, "Addressing the New Reality of Current Visa Policy on International Students and Researchers," testimony, *Hearing Before the U.S. Senate Committee on Foreign Relations*, 108th Cong. (October 6, 2004), 5–7, www.gpo.gov/fdsys/search/home.action.

56. Terry Hartle and Steven Bloom, interview.

57. Heather Stewart, counsel and director of immigration policy, NAFSA, interview by author, Washington, DC, September 11, 2015.

58. Terry Hartle and Steven Bloom, interview.

59. Heather Stewart, interview.

60. David Ward, "Dealing with Foreign Students and Scholars in an Age of Terrorism," 24.

61. Diana Jean Schemo, "Problems Slow Tracking of Students from Abroad," *New York Times*, March 23, 2003.

62. Marlene Johnson, "Addressing the New Reality of Current Visa Policy on International Students and Researchers," testimony, *Hearing Before the U.S. Senate Committee on Foreign Relations*, 108th Cong. (October 6, 2004), 64, www.gpo.gov/fdsys/search/home.action.

63. Federation for American Immigration Reform, "Immigration Issues: Foreign Students," *Federation for American Immigration Reform*, May 2012, www.fairus.org/issue/foreign -students. The president of FAIR, Daniel Stein, later recalled that this position was probably taken in the early 2000s. Daniel Stein, interview.

64. Roy Beck, president, NumbersUSA, interview by author, June 13, 2015.

65. George J. Borjas, "Rethinking Foreign Students," *National Review* 17 (June 17, 2002): 38–41. Borjas later published research finding a negative correlation between increases in foreign students at a particular university and the number of native white male students in the university's graduate program, with the "crowd-out effect" strongest at elite institutions. George J. Borjas, "Do Foreign Students Crowd Out Native Students from Graduate Programs?" Working Paper No. 10349 (Cambridge, MA: National Bureau of Economic Research, 2004), www.nber.org/papers/w10349.

66. George J. Borjas, "An Evaluation of the Foreign Student Program," *Center for Immigration Studies* (Washington, DC: Center for Immigration Studies, June 1, 2002), http://cis.org /ForeignStudentProgram.

67. "Dealing with Foreign Students and Scholars in an Age of Terrorism: Visa Backlogs and Tracking Systems," *Hearing Before the U.S. House of Representatives Committee on Science*, 108th Cong. (March 26, 2003), www.gpo.gov/fdsys/search/home.action; "Addressing the New Reality of Current Visa Policy on International Students and Researchers," *Hearing Before the U.S. Senate Committee on Foreign Relations*, 108th Cong. (October 6, 2004), www .gpo.gov/fdsys/search/home.action.

68. U.S. Department of State, "Nonimmigrant Worldwide Issuance and Refusal Data by Visa Category," *Travel.State.Gov*, January 14, 2014, http://travel.state.gov/content/visas/english /law-and-policy/statistics/non-immigrant-visas.html.

69. U.S. General Accounting Office, *Improvements Needed to Reduce Time Taken to Adjudicate Visas for Science Students and Scholars* (Washington, DC: U.S. General Accounting Office, February 24, 2004), 2, www.gao.gov/products/GAO-04-443T.

70. U.S. Government Accountability Office, *Streamlined Visas Mantis Program Has Lowered Burden on Foreign Science Students and Scholars, but Further Refinements Needed* (Washington, DC: U.S. Government Accountability Office, February 18, 2005), 2, www.gao.gov/products /GAO-05-198.

71. U.S. Government Accountability Office, *Streamlined Visas Mantis Program Has Lowered Burden*, 7.

72. U.S. Government Accountability Office, *Challenges in Attracting International Students to the United States and Implications for Global Competitiveness* (Washington, DC: U.S. Government Accountability Office, June 29, 2007), 13, www.gao.gov/products/GAO-07 -1047T.

73. Marlene Johnson, "Addressing the New Reality of Current Visa Policy on International Students and Researchers," 64.

74. U.S. Government Accountability Office, *Performance of Foreign Student and Exchange Visitor Information System Continues to Improve, but Issues Remain* (Washington, DC: U.S. Government Accountability Office, March 17, 2005), 2, www.gao.gov/products /GAO-05-440T.

75. Bureau of Industry and Security, U.S. Department of Commerce, "Revision and Clarification of Deemed Export Related Regulatory Requirements," *Federal Register*, March 28, 2005.

76. U.S. regulations consider an "export" of technology to take place when it is released to a foreign national within the United States. So, if the technology is controlled in a given case, a license must be issued before the release can take place.

77. Edward Alden and Stephanie Kirchgaessner, "Universities in Fury at Plan to Curb 'Chinese Espionage,'" *Financial Times*, November 25, 2005.

78. Toby Smith, Association of American Universities, and Robert Hardy, Council on Governmental Relations, interview by author, May 23, 2017. I thank Smith and Hardy for providing me with copies of these letters.

79. Danielle Belopotosky, "Policy Change On 'Deemed Exports' Is Widely Panned," *Technology Daily PM*, August 19, 2005.

80. Bureau of Industry and Security, U.S. Department of Commerce, "Revisions and Clarification of Deemed Export Related Regulatory Requirements," *Federal Register*, May 31, 2006, 30840.

81. Robert Hardy, "Commerce Withdraws ANPR on Deemed Exports," May 31, 2006. I thank Hardy for providing me with a copy of this memo.

82. Between 2010 and 2014, for example, the total number of deemed export license applications ranged from 633 to 1,450 per year. The vast majority (92 percent) of these applications were approved, 7 percent were returned without action, and less than 1 percent were denied. Bureau of Industry and Security, U.S. Department of Commerce, response to Freedom of Information Act request, tracking number BIS 15-136, February 11, 2016.

83. Terry Hartle and Steven Bloom, interview.

84. U.S. Department of State, "Nonimmigrant Worldwide Issuance and Refusal Data by Visa Category."

85. Jeff Allum, "Findings from the 2014 CGS International Graduate Admissions Survey—Phase II: Final Applications and Initial Offers of Admission" (Washington, DC: Council of Graduate Schools, 2014), 5.

86. Karin Fischer, "State Department Promises Speedier Visa Review," *Chronicle of Higher Education*, June 12, 2009.

87. Times Higher Education, "World University Rankings 2014–2015," 2014, *Times Higher Education*, www.timeshighereducation.co.uk/world-university-rankings/2014-15/world-ranking.

88. Joseph Carroll, "American Public Opinion About Immigration," *Gallup.com*, July 26, 2005, www.gallup.com/poll/14785/Immigration.aspx.

5. THE (MOSTLY) OPEN DOOR

1. For more background on CFIUS, see James K. Jackson, *The Committee on Foreign Investment in the United States (CFIUS)* (Washington, DC: Congressional Research Service, 2014).

2. Harris Miller, president, Information Technology Association of America, 1995 to 2005, interview by author, July 1, 2015.

3. Robert T. Kudrle and Davis B. Bobrow, "U.S. Policy Toward Foreign Direct Investment," *World Politics* 34, no. 3 (April 1982): 367.

4. Andrew D. Gross and Michael S. Schadewald, "Prospects for U.S. Corporate Tax Reform," *The CPA Journal* 82, no. 1 (January 1, 2012).

5. Gross and Schadewald, "Prospects for U.S. Corporate Tax Reform."

6. "U.S. Led Effort Reaches 'Major Breakthrough' to Expand Information Technology Agreement," *Office of the United States Trade Representative*, July 2015, https://ustr.gov/about-us/policy-offices/press-office/press-releases/2015/july/us-led-effort-reaches-%E2%80%98major.

7. For example, see the recommendations in Daniel Marschall and Laura Clawson, *Sending Jobs Overseas: The Cost to America's Economy and Working Families* (Washington, DC: Working America and the AFL-CIO, 2010).

8. Hugo Meijer, *Trading with the Enemy: The Making of US Export Control Policy Toward the People's Republic of China* (Oxford: Oxford University Press, 2016), 165–97.

9. One technology that remains controlled for both civilian and military users is radiation-hardened microprocessors, since these are frequently used in weapons systems.

10. Christopher A. Padilla, vice-president, government and regulatory affairs, IBM, interview by author, March 25, 2016. Padilla also noted that when export controls have been an issue, IBM has held discussions with U.S. officials in advance concerning the specific technology and the specific end user in order to address any potential concerns.

11. Bill Greenwalt, "We Haven't Won Yet on Export Control Reforms," *Breaking Defense*, November 21, 2013, http://breakingdefense.com/2013/11/we-havent-won-yet-on-export-control-reforms/; Doug Cameron and Julian E. Barnes, "Pentagon Criticizes Contractors' R&D," *Wall Street Journal*, November 21, 2014.

12. Andrew B. Kennedy, "Unequal Partners: U.S. Collaboration with China and India in Research and Development," *Political Science Quarterly* 132, no. 1 (2017), 71.

13. Kennedy, "Unequal Partners," 72.

14. Quoted in Steve Lohr, "New Economy: Offshore Jobs in Technology; Opportunity or a Threat?" *New York Times*, December 22, 2003, www.nytimes.com/2003/12/22/business/new-economy-offshore-jobs-in-technology-opportunity-or-a-threat.html.

15. Quoted in Steve Lohr, "Many New Causes for Old Problem of Jobs Lost Abroad," *New York Times*, February 15, 2004.

16. Ron Hira, "Implications of Offshore Outsourcing" (Paper presented at the Globalization, Employment, and Economic Development Workshop, Sloan Workshop Series in Industry Studies, Rockport, MA, January 3, 2004), 4–5.

17. Alec Gallup and Frank Newport, *The Gallup Poll: Public Opinion 2004* (Lanham, MD: Rowman and Littlefield, 2006), 110.

18. Joseph I. Lieberman, *Offshore Outsourcing and America's Competitive Edge: Losing Out in the High Technology R&D and Services Sectors* (Washington, DC: Office of Senator Joseph I. Lieberman, May 11, 2004), 5.

19. Lieberman, "Offshore Outsourcing and America's Competitive Edge," 16.

20. William Reinsch, "What Is to Be Done on Trade?" *Stimson Spotlight*, June 7, 2016, www.stimson.org/content/what-be-done-trade. Reinsch was president of the National Foreign Trade Council (NFTC) from 2001 to 2016, when he also served as a member of the U.S.–China Economic and Security Review Commission. Prior to joining the NFTC, Reinsch served as the under-secretary for export administration in the U.S. Department of Commerce and as a legislative assistant on Capitol Hill.

21. Jonathan Weisman, "Bush, Adviser Assailed for Stance on 'Offshoring' Jobs," *Washington Post*, February 11, 2004.

22. Alan S. Blinder, "Offshoring: The Next Industrial Revolution?" *Foreign Affairs* 85, no. 2 (2006): 113–28.

23. J. Bradford Jensen, *Global Trade in Services: Fear, Facts, and Offshoring* (Washington, DC: Peterson Institute for International Economics, 2011).

24. Gary Clyde Hufbauer, Theodore H. Moran, and Lindsay Oldenski, *Outward Foreign Direct Investment and US Exports, Jobs, and R&D: Implications for US Policy* (Washington, DC: Peterson Institute for International Economics, 2013).

25. "IEEE Advocates Limits on Offshore Outsourcing," *Information Week*, March 15, 2004, www.informationweek.com/ieee-advocates-limits-on-offshore-outsourcing/d/d-id/1023812.

26. Melissa Block, "Analysis: Industry Groups Fight Anti-outsourcing Legislation," *National Public Radio: All Things Considered*, March 5, 2004.

27. Michael Schroeder, "Business Coalition Battles Outsourcing Backlash," *Wall Street Journal*, March 1, 2004, www.wsj.com/articles/SB107809268846542227.

28. Schroeder, "Business Coalition Battles Outsourcing Backlash."

29. Block, "Analysis: Industry Groups Fight Anti-outsourcing Legislation."

30. Ben Worthen, "Regulations: What to Worry About," *CIO*, June 15, 2004.

31. American Federation of Labor and Congress of Industrial Organizations (AFL-CIO), "Summary of Activities Regarding Off-Shore Outsourcing, June 2003–May 2004," *Department for Professional Employees, AFL-CIO,* 2004, http://dpeaflcio.org/archives /legislative-reports-2/summary-of-activities-regarding-off-shore-outsourcing-june -2003-may-2004/.

32. An exception was allowed for work that was already being done outside the United States. See Wilson Dizard, "New Law Will Curb Offshoring of Federal IT Work," *Government Computer News,* February 23, 2004.

33. Harris Miller, interview.

34. Quoted in Shumita Sharma, "US Offshore Outsourcing Ban Sparks Fears of Similar Laws," *Dow Jones International Newswires,* January 30, 2004.

35. Harris Miller, interview.

36. AFL-CIO, "Summary of Activities Regarding Off-Shore Outsourcing."

37. Martin Vaughan and Susan Davis, "Senate Ends Corporate Tax Debate for Now, OKs Outsourcing Deal," *Congress Daily,* March 5, 2004.

38. Lori Simpson, *Engineering Aspects of Offshore Outsourcing* (Alexandria, VA: National Society of Professional Engineers, August 6, 2004), 3, www.wise-intern.org/journal/2004 /wise2004-lorisimpsonfinalpaper.pdf.

39. Whereas Dodd's initial bill had allowed national security exceptions for contracts certified by the president, the changes gave agency heads broad authority to determine which contracts were for purposes of national security. In addition, the new language made the restrictions contingent on annual certification by the Department of Commerce that U.S. job losses resulting from the restrictions were not greater than the number of jobs they preserved. Last, the new language exempted the Departments of Defense and Homeland Security, intelligence agencies, and the national security programs of the Department of Energy. See Vaughan and Davis, "Senate Ends Corporate Tax Debate for Now, OKs Outsourcing Deal."

40. Jobs and Trade Network, "Jobs and Trade Network to Hold Press Luncheon with U.S. Sen. Dodd: National Fair Trade Group to Advocate Against Outsourcing Policies," *Economic Policy Institute,* February 20, 2004.

41. Brian Tumulty, "Small Manufacturers Aim Buy American Challenge at U.S. Job Losses," *Gannett News Service,* October 31, 2003, http://global.factiva.com/redir/default.aspx?P=sa&an =GNS0000020040107dzavoooj7&cat=a&ep=ASE.

42. Ed Frauenheim, "Tech Professionals Group Wary of Offshoring," *CNET News,* March 18, 2004.

43. Harris Miller, interview.

44. Harris Miller, interview.

45. Keith Koffler, "Business Coalition Rewrites Lexicon for Jobs 'Outsourcing,'" *Congress Daily,* March 2, 2004.

46. Gail Repsher Emery, "Industry Fights Dodd Legislation," *Newsbytes News Network,* April 15, 2004.

47. Repsher Emery, "Industry Fights Dodd Legislation."

48. Mary Dalrymple, "Senate Passes Corporate Tax Bill," *Associated Press*, May 12, 2004.

49. "US's Grassley—'Difficult' to Get Quick Action on Tax Bill," *Market News International*, June 22, 2004.

50. Jonathan Weisman and Mark Kaufman, "Tax-Cut Bill Draws White House Doubts: Corporate Provisions Go Beyond 'Core Objective,' Treasury Secretary Says," *Washington Post*, October 5, 2004.

51. Paul Almeida, president, AFL-CIO Department for Professional Employees, interview by author, August 21, 2015.

52. AFL-CIO, "Outsourcing America," *AFL-CIO*, March 11, 2004, www.aflcio.org/About /Exec-Council/EC-Statements/Outsourcing-America.

53. "Kerry Tax Plan Proposes to Slow Loss of US Jobs Overseas," *Dow Jones International News*, March 26, 2004.

54. David J. Lynch, "Does Tax Code Encourage U.S. Companies to Cut Jobs at Home? Presidential Candidates Target Corporate Tax Breaks for Offshoring," *USA Today*, March 21, 2008.

55. Barack Obama, "Remarks of President Barack Obama: Address to Joint Session of Congress," *Whitehouse.gov*, February 24, 2009, www.whitehouse.gov/the-press-office /remarks-president-barack-obama-address-joint-session-congress.

56. "Obama Lowers Temperature Against Outsourcing," *Economic Times*, March 28, 2009.

57. Office of the Press Secretary, "Leveling the Playing Field: Curbing Tax Havens and Removing Tax Incentives for Shifting Jobs Overseas," *U.S. Department of the Treasury Press Center*, May 4, 2009, www.treasury.gov/press-center/press-releases/Pages/tg119.aspx.

58. Jackie Calmes and Edmund Andrews, "Obama Seeks to Curb Foreign Tax Havens," *New York Times*, May 4, 2009, www.nytimes.com/2009/05/05/business/05tax.html?pagewanted =all&_r=1&.

59. Kate Ackley, "Offshore Accounts Present a Taxing Situation," *Roll Call*, May 6, 2009; "Bond, Thune Spearhead Efforts to Gut Foreign Tax Provisions from Bill," *Inside U.S. Trade*, June 18, 2010.

60. Neil King and Elizabeth Williamson, "Business Fends Off Tax Hit: Obama Administration Shelves Plan to Change How U.S. Treats Overseas Profits," *Wall Street Journal*, October 13, 2009.

61. King and Williamson, "Business Fends Off Tax Hit."

62. Obama administration White House official, interview by author, April 14, 2017.

63. Obama administration White House official, interview.

64. John D. McKinnon, "Plan Would Raise Taxes on Businesses," *Wall Street Journal*, February 2, 2010, www.wsj.com/articles/SB10001424052748704107204575039073372259004.

65. Shailagh Murray, "In Senate and on Trail, Democrats Target Jobs Moving Abroad," *Washington Post*, June 9, 2010.

66. Richard Trumka, "Statement by AFL-CIO President Richard Trumka on the Promoting American Jobs and Closing Tax Loopholes Act," *AFL-CIO*, May 24, 2010, http://ftp .workingamerica.org/Press-Room/Press-Releases/Statement-by-AFL-CIO -President-Richard-Trumka-on-t14.

67. Quoted in Murray, "In Senate and on Trail, Democrats Target Jobs Moving Abroad."

68. "Bond, Thune Spearhead Efforts to Gut Foreign Tax Provisions from Bill."

69. Peter Schroeder, "Extender Efforts Hit Roadblock as Senate Tables Tax Package," *Bond Buyer*, June 28, 2010.

70. "Amendments Submitted and Proposed," *Congressional Record*, 111th Cong. 156, no. 115 (August 2, 2010): S6586–93.

71. Martin Vaughan, "Businesses Split Over Tax Credits," *Wall Street Journal*, August 4, 2010.

72. "The Eternal 'Emergency,'" *Wall Street Journal*, August 6, 2010.

73. FAA Air Transportation Modernization and Safety Improvement Act of 2010, H.R. 1586, 111th Cong. (2010), www.govtrack.us/congress/bills/111/hr1586.

74. Ronald Brownstein, "Back to Basics," *National Journal*, September 9, 2010.

75. Creating American Jobs and Ending Offshoring Act of 2010, S.3816, 111th Cong. (2010), www.govtrack.us/congress/bills/111/s3816.

76. Philip J. Bond, "Opposition to S.3816," September 27, 2010, www.techamerica.org/content /wp-content/uploads/2010/09/9-27-10_S3816.pdf.

77. Jennifer Liberto and Dana Bash, "Bill to Hike Taxes on Overseas Jobs Fails Senate Test Vote," *CNN*, September 28, 2010, http://money.cnn.com/2010/09/28/news/economy /Outsource_jobs_bill_dead/.

78. Richard Trumka, "Statement by AFL-CIO President Richard Trumka on Creating American Jobs and Ending Offshoring Act," *AFL-CIO*, September 28, 2010, www.aflcio.org /Press-Room/Press-Releases/Statement-by-AFL-CIO-President-Richard-Trumka-on-C8.

79. Melanie Trottman, "Web Tool Could Help Boost Union Voter Turnout," *Wall Street Journal*, October 7, 2010.

80. Marschall and Clawson, "Sending Jobs Overseas: The Cost to America's Economy and Working Families," 22.

81. Daniel Marschall, legislative and policy specialist for workforce issues, AFL-CIO, interview by author, December 18, 2014. Marschall was offering his personal view, rather than speaking as a spokesperson for the AFL-CIO.

82. Patrick Thibodeau, interview by author, September 8, 2015. Thibodeau has covered the offshoring issue for more than a decade as a Washington-based correspondent for *Computerworld*.

83. Bring Jobs Home Act of 2011/12, S.3364, 112th Cong. (2011/12), www.congress.gov/bill /112th-congress/senate-bill/3364.

84. Bring Jobs Home Act of 2013/14, S.2569, 113th Cong. (2013/14), www.congress.gov/bill /112th-congress/senate-bill/2569.

85. National Foundation for American Policy, *NFAP Policy Brief: Anti-outsourcing Efforts Down but Not Out* (Arlington, VA: National Foundation for American Policy, April 2007), 2, 4, www.nfap.com/pdf/0407OutsourcingBrief.pdf.

86. Greg Schneider, "Anxious About Outsourcing: States Try to Stop U.S. Firms from Sending High-Tech Work Overseas," *Washington Post*, January 31, 2004.

87. Quoted in Carolyn Duffy Marsan, "A Political Hot Potato: Legislatures Juggle Offshore Outsourcing Regulations," *Network World*, July 5, 2004.

88. Quoted in Block, "Analysis: Industry Groups Fight Anti-outsourcing Legislation."

89. National Foundation for American Policy, "Anti-Outsourcing Efforts Down but Not Out," 2–3.

90. Michael Schroeder, "States' Efforts to Curb Outsourcing Stymied: Business Groups Take the Lead in Weakening Attempts to Limit Work from Moving Abroad," *Wall Street Journal*, April 16, 2004.

91. Quoted in Schroeder, "States' Efforts to Curb Outsourcing Stymied."

92. Harris Miller, interview.

93. The law allowed exemptions if a service could not be performed within the United States or if application of the law would violate the terms of any grant, funding, or financial assistance from the federal government. See Service Contract Requirements for the Performance of Service Contracts Within the United States, Pub. L. No. 2005, c. 92 (New Jersey, 2005), www.njleg.state.nj.us/bills/BillView.asp.

94. Michael Schroeder, "States Fight Exodus of Jobs: Lawmakers, Unions Seek to Block Outsourcing Overseas," *Wall Street Journal*, June 3, 2003.

95. Bill Tucker, "Job Creation Stalls: Interview with Commerce Secretary Don Evans," *Lou Dobbs Tonight* (CNN, January 9, 2004).

96. Schroeder, "States Fight Exodus of Jobs."

97. Schneider, "Anxious About Outsourcing."

98. Quoted in Schneider, "Anxious About Outsourcing."

99. James E. McGreevey, *The Confession* (New York: Harper Collins, 2006), 223.

100. Quoted in "McGreevey Edict Restricts Outsourcing by Agencies," *Congress Daily*, September 13, 2004.

101. Shirley Turner, New Jersey state senator, interview by author, September 11, 2015.

102. David Kocienieski, "For Trenton's Lame Duck, the Question Is, 'How Lame?,'" *New York Times*, September 12, 2004.

103. "McGreevey Edict Restricts Outsourcing by Agencies."

104. Quoted in Josh Gohlke, "Cipel Was No Security Risk, McGreevey Says: Governor Speaks of Former Aide by Name for First Time," *Record*, September 10, 2004.

105. Shirley Turner, interview.

106. Shirley Turner, interview.

107. "New Jersey Governor Quits, Comes Out as Gay," *CNN*, August 13, 2004, http://edition.cnn.com/2004/ALLPOLITICS/08/12/mcgreevey.nj/.

108. LexisNexis State Net database, accessed August 14, 2015.

109. Patrick Thibodeau, "Ohio Bans Offshoring as It Gives Tax Relief to Outsourcing Firm," *Computerworld*, September 7, 2010, www.computerworld.com/article/2515465/it-outsourcing/ohio-bans-offshoring-as-it-gives-tax-relief-to-outsourcing-firm.html.

110. The number of U.S. articles in the Factiva database containing the word *offshoring* or *outsourcing* and the words *states* and *bills* peaks at 3,441 for 2004 and declines to less than 2,000 per year thereafter (except for upticks in 2010 and 2012, when the totals reached 2,117 and 2,504, respectively).

CONCLUSION

1. Robert O. Keohane, "The Old IPE and the New," *Review of International Political Economy* 16, no. 1 (2009): 34.
2. Edward D. Mansfield and Diana C. Mutz, "US Versus Them: Mass Attitudes Toward Offshore Outsourcing," *World Politics* 65, no. 4 (2013): 571–608.
3. Brenton D. Peterson, Sonal S. Pandya, and David Leblang, "Doctors with Borders: Occupational Licensing as an Implicit Barrier to High Skill Migration," *Public Choice* 160, no. 1–2 (July 2014): 45–63; Neil Malhotra, Yotam Margalit, and Cecilia Hyunjung Mo, "Economic Explanations for Opposition to Immigration: Distinguishing Between Prevalence and Conditional Impact," *American Journal of Political Science* 57, no. 2 (April 2013): 391–410; Giovanni Facchini and Anna Maria Mayda, "Individual Attitudes Towards Skilled Migration: An Empirical Analysis Across Countries," *The World Economy* 35, no. 2 (2012): 183–96; Jens Hainmueller and Michael J. Hiscox, "Attitudes Toward Highly Skilled and Low-Skilled Immigration: Evidence from a Survey Experiment," *American Political Science Review* 104, no. 1 (February 2010): 61–84; Kenneth F. Scheve and Matthew Jon Slaughter, *Globalization and the Perceptions of American Workers* (Washington, DC: Peterson Institute, 2001).
4. Peterson, Pandya, and Leblang, "Doctors with Borders."
5. Margaret E. Peters, "Trade, Foreign Direct Investment, and Immigration Policy Making in the United States," *International Organization* 68, no. 4 (2014): 811–44.
6. Ayelet Shachar, "Talent Matters: Immigration Policy-Setting as a Competitive Scramble Among Jurisdictions," in *Wanted and Welcome? Policies for Highly Skilled Immigrants in Comparative Perspective*, ed. Triadafilos Triadafilopoulos (New York: Springer, 2013), 91.
7. William Aspray, Frank Mayadas, and Moshe Vardi, "Globalization and Offshoring of Software: A Report of the ACM Job Migration Task Force" (New York: Association for Computing Machinery, 2006), 52, www.acm.org/globalizationreport.
8. For example, see Martin Ruhs, *The Price of Rights: Regulating International Labor Migration* (Princeton, NJ: Princeton University Press, 2013). Gary Freeman also appears to incline toward this view, at least with regard to European countries. See Gary P. Freeman, "National Models, Policy Types, and the Politics of Immigration in Liberal Democracies," *West European Politics* 29, no. 2 (March 1, 2006): 237–38.
9. Shachar, "Talent Matters"; Christiane Kuptsch and Eng Fong Pang, "Introduction," in *Competing for Global Talent*, ed. Christiane Kuptsch and Eng Fong Pang (Geneva: International Labor Office, 2006), 1–8; James F. Hollifield, "The Emerging Migration State," *International Migration Review* 38, no. 3 (2004): 191.
10. Georg Menz, *The Political Economy of Managed Migration: Nonstate Actors, Europeanization, and the Politics of Designing Migration Policies* (Oxford: Oxford University Press, 2008); Lucie Cerna, "The Varieties of High-Skilled Immigration Policies: Coalitions and Policy Outputs in Advanced Industrial Countries," *Journal of European Public Policy* 16, no. 1 (January 2009): 144–61; Alexander A. Caviedes, *Prying Open Fortress Europe: The Turn to Sectoral Labor Migration* (Lanham, MD: Lexington Books, 2010); Lucie Cerna, "Attracting

High-Skilled Immigrants: Policies in Comparative Perspective," *International Migration* 52, no. 3 (June 2014): 69–84.

11. Chris F. Wright, "Why Do States Adopt Liberal Immigration Policies? The Policymaking Dynamics of Skilled Visa Reform in Australia," *Journal of Ethnic and Migration Studies* 41, no. 2 (January 28, 2015): 306–28.

12. Hollifield, "The Emerging Migration State."

13. Linsu Kim, *Imitation to Innovation: The Dynamics of Korea's Technological Learning* (Cambridge, MA: Harvard Business Press, 1997); Sean O'Riain, *The Politics of High Tech Growth: Developmental Network States in the Global Economy* (Cambridge: Cambridge University Press, 2004); Dan Breznitz, *Innovation and the State* (New Haven, CT: Yale University Press, 2007); Dieter Ernst, *A New Geography of Knowledge in the Electronics Industry? Asia's Role in Global Innovation Networks* (Honolulu, HI: East-West Center, 2009); Joseph Wong, *Betting on Biotech: Innovation and the Limits of Asia's Developmental State* (New York: Cornell University Press, 2011); Sung-Young Kim, "Transitioning from Fast-Follower to Innovator: The Institutional Foundations of the Korean Telecommunications Sector," *Review of International Political Economy* 19, no. 1 (February 2012): 140–68.

14. Andrew B. Kennedy, "Slouching Tiger, Roaring Dragon: Comparing India and China as Late Innovators," *Review of International Political Economy* 23, no. 2 (2016): 1–28.

15. The literature on techno-nationalism and techno-globalism in developing countries represents a partial exception to this tendency, but this body of work has been geographically limited, focusing mainly on East Asian cases, and is more concerned with technological development than with innovation in particular. For a review of this literature, see Andrew B. Kennedy, "China's Search for Renewable Energy: Pragmatic Techno-Nationalism," *Asian Survey* 53, no. 5 (2013): 911–13.

16. Mancur Olson, *The Rise and Decline of Nations: Economic Growth, Stagflation, and Social Rigidities* (New Haven, CT: Yale University Press, 1982), 217.

17. Espen Moe, *Governance, Growth and Global Leadership: The Role of the State in Technological Progress, 1750–2000* (Hampshire, UK: Ashgate, 2013); Mark Zachary Taylor, *The Politics of Innovation: Why Some Countries Are Better Than Others at Science and Technology* (Oxford: Oxford University Press, 2016).

18. Michael G. Finn, "Stay Rates of Foreign Doctorate Recipients from U.S. Universities, 2001" (Oak Ridge, TN: Oak Ridge Institute for Science and Education, November 2003), 7, http://orise.orau.gov/files/sep/stay-rates-foreign-doctorate-recipients-2001.pdf.

19. Testimony of Bruce Morrison, *Hearing Before the Committee on the Judiciary of the United States House of Representatives Subcommittee on Immigration Policy and Enforcement*, March 5, 2013, 4.

20. Vivek Wadhwa, *The Immigrant Exodus: Why America Is Losing the Global Race to Capture Entrepreneurial Talent* (Philadelphia: Wharton Digital, 2012), 49.

21. "Four Ways to Tackle H-1B Visa Reform," *IEEE Spectrum*, April 19, 2017.

22. David Bier, "No One Knows How Long Legal Immigrants Will Have to Wait," *Cato Institute*, July 28, 2016, www.cato.org/blog/no-one-knows-how-long-legal-immigrants -will-have-wait.

23. Andrew B. Kennedy, "Unequal Partners: U.S. Collaboration with China and India in Research and Development," *Political Science Quarterly* 132, no. 1 (2017): 63–86.

24. Lee Branstetter, Guangwei Lee, and Francisco Veloso, "The Rise of International Coinvention," in *The Changing Frontier: Rethinking Science and Innovation Policy*, ed. Adam B. Jaffe and Benjamin F. Jones (Chicago: University of Chicago Press, 2015), 140, 159.

25. This finding was attributed to several factors, including increased competition for local talent, zealous efforts by foreign firms to protect intellectual property, the gap between foreign and local capabilities, and the tendency of foreign firms to do core R&D in their home countries. See Xiaolan Fu and Yundan Gong, "Indigenous and Foreign Innovation Efforts and Drivers of Technological Upgrading: Evidence from China," *World Development* 39, no. 7 (July 2011): 1213–25.

26. Xiaohong Quan, "Knowledge Diffusion from MNC R&D Labs in Developing Countries: Evidence from Interaction Between MNC R&D Labs and Local Universities in Beijing," *International Journal of Technology Management* 51, no. 2 (2010): 364–86.

27. Anabel Marin and Subash Sasidharan, "Heterogeneous MNC Subsidiaries and Technological Spillovers: Explaining Positive and Negative Effects in India," *Research Policy* 39, no. 9 (November 2010): 1227–41.

28. Kennedy, "Slouching Tiger, Roaring Dragon."

29. Mark Foulon and Christopher A. Padilla, "In Pursuit of Security and Prosperity: Technology Controls for a New Era," *Washington Quarterly* 30, no. 2 (2007): 88.

30. Mark Foulon, former acting undersecretary of Commerce for Industry and Security, interview by author, July 21, 2015.

31. Christopher A. Padilla, vice-president, Government and Regulatory Affairs, IBM, interview by author, March 25, 2016.

32. In general, a potential obstacle to greater information-sharing between government and the private sector is the private sector's fear of inadvertently disclosing what government enforcement officials might consider a violation of U.S. export controls. For greater information-sharing to occur, the U.S. government will have to find a way to reassure companies in this regard without abandoning its overall commitment to export control. William Reinsch, president, National Foreign Trade Council, interview by author, February 3, 2016.

33. Vernon M. Briggs, *Immigration and American Unionism* (Ithaca, NY: Cornell University Press, 2001); Aristide Zolberg, *A Nation by Design: Immigration Policy in the Fashioning of America* (Cambridge, MA: Harvard University Press, 2006); Daniel J. Tichenor, *Dividing Lines: The Politics of Immigration Control in America* (Princeton, NJ: Princeton University Press, 2002); Peters, "Trade, Foreign Direct Investment, and Immigration Policy Making in the United States."

34. Jennifer Steinhauer, Jonathan Martin, and David M. Herszenhorn, "Paul Ryan Calls Donald Trump's Attack on Judge 'Racist,' but Still Backs Him," *New York Times*, June 7, 2016, www.nytimes.com/2016/06/08/us/politics/paul-ryan-donald-trump-gonzalo-curiel.html.

35. Seth Fiegerman, "Silicon Valley Throws Big Money at Clinton and Virtually Nothing at Trump," *CNN*, August 23, 2016, http://money.cnn.com/2016/08/23/technology/hillary-clinton-tech-fundraisers/.

36. Jessica Meyers, "Tech Companies See Few Big Gains in Obama's Executive Action," *Boston Globe*, November 24, 2014, www.bostonglobe.com/news/nation/2014/11/24/tech-companies-see-few-big-gains-obama-executive-action/dauDJujkOhe1qx5ZQTScoM/story.html.

37. Todd Frankel and Tracy Jan, "Trump's New Travel Ban Raises the Same Silicon Valley Objections," *Washington Post*, March 6, 2017, www.washingtonpost.com/news/the-switch/wp/2017/03/06/trumps-new-travel-ban-raises-the-same-silicon-valley-objections/.

38. Rebecca Dickson, "Trump Officials Clamp Down on Worker Visas," *The Hill*, April 6, 2017, http://thehill.com/business-a-lobbying/business-a-lobbying/327507-trump-officials-clamp-down-on-worker-visas.

39. Tony Romm, "How Silicon Valley Is Trying to Topple Trump—Beginning with a Special Election in Montana," *Recode*, May 25, 2017, www.recode.net/2017/5/25/15686802/silicon-valley-trump-montana-tech-for-campaigns.

40. Donald J. Trump, "Presidential Executive Order on Buy American and Hire American," *White House*, April 18, 2017, www.whitehouse.gov/the-press-office/2017/04/18/presidential-executive-order-buy-american-and-hire-american.

41. Joshua Brustein, "The Secret Way Silicon Valley Uses the H-1B Program," *Bloomberg*, June 6, 2017, www.bloomberg.com/news/articles/2017-06-06/silicon-valley-s-h-1b-secret.

42. Kate Kelly, Rachel Abrams, and Alan Rappeport, "Trump Is Said to Abandon Contentious Border Tax on Imports," *New York Times*, April 25, 2017, www.nytimes.com/2017/04/25/us/politics/orrin-hatch-trump-tax-cuts-deficit-economy.html.

43. "Briefing by Secretary of the Treasury Steven Mnuchin and Director of the National Economic Council Gary Cohn," *White House*, April 26, 2017, www.whitehouse.gov/the-press-office/2017/04/26/briefing-secretary-treasury-steven-mnuchin-and-director-national.

44. Information Technology Industry Council, "ITI on Trump Tax Reform Principles," *Information Technology Industry Council*, April 26, 2017, www.itic.org/news-events/news-releases/iti-on-trump-tax-reform-principles; Tony Romm, "Finally, Silicon Valley and Donald Trump Agree on Something: Taxes," *Recode*, April 26, 2017, www.recode.net/2017/4/26/15437330/silicon-valley-tech-donald-trump-agree-tax-repatriation-reform.

45. Joseph G. Paul and Frank Caruso, "One of Trump's Biggest Plans to Stimulate the Economy Won't Be Great for Most Americans," *Business Insider*, June 1, 2017, www.businessinsider.com.au/alliancebernstein-on-trump-tax-plan-2017-5?r=US&IR=T.

BIBLIOGRAPHY

Abate, Tom, and Jon Swartz. "Eleventh-Hour Victory for Tech: Visa Increase, R&D Tax Measure in Budget Bill." *San Francisco Chronicle*, October 16, 1998. www.sfgate.com/business /article/11th-Hour-Victory-For-Tech-Visa-increase-R-D-2984825.php.

"About SMART." *Singapore–MIT Alliance for Research and Technology*, 2013. http://smart.mit .edu/about-smart/about-smart.html.

Ackley, Kate. "Offshore Accounts Present a Taxing Situation." *Roll Call*, May 6, 2009.

"Address by House Speaker J. Dennis Hastert: Reflections on Role of Speaker in Modern Day House of Representatives." *U.S. Newswire*, November 12, 2003.

"Addressing the New Reality of Current Visa Policy on International Students and Researchers." *Hearing Before the U.S. Senate Committee on Foreign Relations*. 108th Cong., October 6, 2004. www.gpo.gov/fdsys/search/home.action.

Albanesius, Chloe. "Outsourcing Controversy Influences Debate on H-1B Visas." *Technology Daily*, May 6, 2004.

Alden, Edward. "Emigrants to US Face Long Wait for Green Card." *Financial Times*, October 12, 2005.

Alden, Edward, and Stephanie Kirchgaessner. "Universities in Fury at Plan to Curb 'Chinese Espionage.'" *Financial Times*, November 25, 2005.

Allum, Jeff. "Findings from the 2014 CGS International Graduate Admissions Survey—Phase II: Final Applications and Initial Offers of Admission." Washington, DC: Council of Graduate Schools, 2014.

Altbach, Philip G., and Jane Knight. "The Internationalization of Higher Education: Motivations and Realities." *Journal of Studies in International Education* 11, no. 3–4 (2007): 290–305.

Altbach, Philip G., and Patti McGill Peterson. "Internationalize American Higher Education? Not Exactly." *Change* 30, no. 4 (1998): 36–39.

"Amendments Submitted and Proposed." *Congressional Record*. 111th Cong. 156, no. 115, August 2, 2010.

American Competitiveness in the Twenty-First Century Act of 2000. S.2045. 106th Cong., 2000. www.gpo.gov/fdsys/pkg/BILLS-106s2045enr/pdf/BILLS-106s2045enr.pdf.

American Council on Education. "Higher Education Associations Strongly Endorse Senate Immigration Reform Bill." Letter to the U.S. Senate, June 26, 2013. https://www.aau.edu /sites/default/files/AAU%20Files/Key%20Issues/Budget%20%26%20Appropriations/FY17 /Endorsement-Letter-S-744_6-26-13.pdf.

American Enterprise Institute. "China Global Investment Tracker." *AEI*, April 30, 2017. www.aei .org/china-global-investment-tracker/.

American Federation of Labor and Congress of Industrial Organizations (AFL-CIO). "Immigration." *AFL-CIO*, February 16, 2000. www.aflcio.org/About/Exec-Council/EC -Statements/Immigration2.

——. "Outsourcing America." *AFL-CIO*, March 11, 2004. www.aflcio.org/About/Exec-Council /EC-Statements/Outsourcing-America.

——. "Summary of Activities Regarding Off-Shore Outsourcing, June 2003–May 2004." *Department for Professional Employees, AFL-CIO*, 2004. http://dpeaflcio.org/archives/legislative -reports-2/summary-of-activities-regarding-off-shore-outsourcing-june-2003-may-2004/.

American Immigration Lawyers Association. "Analysis of the American Competitiveness in the Twenty-First Century Act." *Shusterman.com*. Accessed March 2, 2016. http://shusterman .com/h1b-analysisofac21.html.

"Analysis: Look at the Controversy Over H-1B Visas." *National Public Radio: Talk of the Nation*, September 26, 2000.

Archibugi, Daniele, and Simona Iammarino. "The Globalization of Technological Innovation: Definition and Evidence." *Review of International Political Economy* 9, no. 1 (Spring 2002): 98–122.

Aronson, Jonathan D. "International Intellectual Property Rights in a Networked World." In *Power, Interdependence, and Nonstate Actors in World Politics*, edited by Helen V. Milner and Andrew Moravcsik, 185–203. Princeton, NJ: Princeton University Press, 2011.

Arslan, Cansin, Jean-Christophe Dumont, Zovanga Kone, Yasser Moullan, Caglar Ozden, Christopher Parsons, and Theodora Xenogiani. *A New Profile of Migrants in the Aftermath of the Recent Economic Crisis*. OECD Social, Employment and Migration Working Paper No. 160, 2014. www.oecd.org/els/mig/WP160.pdf.

Asheim, Bjørn T., and Meric S. Gertler. "The Geography of Innovation: Regional Innovation Systems." In *The Oxford Handbook of Innovation*, edited by Jan Fagerberg, David C. Mowery, and Richard R. Nelson, 291–317. Oxford: Oxford University Press, 2005.

Asian Development Bank. *Counting the Cost: Financing Asian Higher Education for Inclusive Growth*. Manila: Asian Development Bank, 2012.

Aspray, William, Frank Mayadas, and Moshe Vardi. "Globalization and Offshoring of Software: A Report of the ACM Job Migration Task Force." New York: Association for Computing Machinery, 2006. www.acm.org/globalizationreport.

Australian Government Department of Education. "International Student Enrolments by Nationality in 2013," April 2014. https://internationaleducation.gov.au/research/research-snapshots/pages/default.aspx.

Australian Government Department of Education and Training. "International Students Studying Science, Technology, Engineering and Mathematics (STEM) in Australian Higher Education Institutions." *Research Snapshots*, October 2015. https://internationaleducation.gov.au/research/Research-Snapshots/Documents/STEM%202014.pdf.

Babington, Charles. "Immigration Bill Expected to Pass Senate This Week: Hastert May Block Version That Divides House GOP." *Washington Post*, May 23, 2006.

Ball, Molly. "Immigration Reformers Are Winning August." *Atlantic*, August 21, 2013. www.theatlantic.com/politics/archive/2013/08/immigration-reformers-are-winning-august/278873/.

——. "The Little Group Behind the Big Fight to Stop Immigration Reform." *Atlantic*, August 1, 2013. www.theatlantic.com/politics/archive/2013/08/the-little-group-behind-the-big-fight-to-stop-immigration-reform/278252/.

Baumgartner, Frank R., Jeffrey M. Berry, Marie Hojnacki, Beth L. Leech, and David C. Kimball. *Lobbying and Policy Change: Who Wins, Who Loses, and Why*. Chicago: University of Chicago Press, 2009.

Baumgartner, Frank R., and Beth L. Leech. *Basic Interests: The Importance of Groups in Politics and in Political Science*. Princeton, NJ: Princeton University Press, 1998.

Belopotosky, Danielle. "Efforts to Change Visa Law for Skilled Workers Gains Steam." *Technology Daily*, October 1, 2004.

——. "Lobbyists Push Congress for Action on H-1B Visas." *Technology Daily*, November 16, 2004.

——. "Policy Change On 'Deemed Exports' Is Widely Panned." *Technology Daily PM*, August 19, 2005.

Bernstein, Nina. "In the Streets, Suddenly, an Immigrant Groundswell." *New York Times*, March 27, 2006.

Berry, Jeffrey M. *The New Liberalism: The Rising Power of Citizen Groups*. Washington, DC: Brookings Institution, 1999.

Bier, David. "High-Skilled Immigration Restrictions Are Economically Senseless." *Forbes*, July 22, 2012. www.forbes.com/sites/realspin/2012/07/22/high-skilled-immigration-restrictions-are-economically-senseless/.

——. "No One Knows How Long Legal Immigrants Will Have to Wait." *Cato Institute*, July 28, 2016. www.cato.org/blog/no-one-knows-how-long-legal-immigrants-will-have-wait.

——. "Why Does the Government Care Where Immigrant Workers Were Born?" *Cato Institute*, January 18, 2017. www.cato.org/blog/why-does-government-care-where-immigrant-workers-were-born.

Blinder, Alan S. "Offshoring: The Next Industrial Revolution?" *Foreign Affairs* 85, no. 2 (2006): 113–28.

Block, Melissa. "Analysis: Industry Groups Fight Anti-outsourcing Legislation." *National Public Radio: All Things Considered*, March 5, 2004.

Bloom, Nicholas, Raffaella Sadun, and John Van Reenen. "Americans Do IT Better: US Multinationals and the Productivity Miracle." *The American Economic Review* 102, no. 1 (2012): 167–201.

Bond, Philip J. "Opposition to S.3816." *TechAmerica*, September 27, 2010. www.techamerica.org /content/wp-content/uploads/2010/09/9-27-10_S3816.pdf.

"Bond, Thune Spearhead Efforts to Gut Foreign Tax Provisions from Bill." *Inside U.S. Trade*, June 18, 2010.

Border Security, Economic Opportunity, and Immigration Modernization Act of 2013. S. 744. 113th Cong., 2013. www.govtrack.us/congress/bills/113/s744.

Borjas, George J. "An Evaluation of the Foreign Student Program." Washington, DC: Center for Immigration Studies, June 2002. http://cis.org/ForeignStudentProgram.

——. *Immigration Economics*. Cambridge, MA: Harvard University Press, 2014.

——. "Rethinking Foreign Students." *National Review* 17 (June 17, 2002): 38–41.

Borroz, Tony. "Chevrolet's Mouse that Roared." *Wired*, August 22, 2011. www.wired.com/2011/08 /chevrolets-mouse-that-roared/.

Boyd, Monica. "Recruiting High Skill Labour in North America: Policies, Outcomes and Futures." *International Migration* 52, no. 3 (June 2014): 40–54.

Brandt, Loren, and Eric Thun. "The Fight for the Middle: Upgrading, Competition, and Industrial Development in China." *World Development* 38, no. 11 (November 2010): 1555–74.

Branigin, William. "Visa Deal for Computer Programmers Angers Labor Groups." *Washington Post*, September 27, 1998.

Branstetter, Lee, Guangwei Lee, and Francisco Veloso. "The Rise of International Coinvention." In *The Changing Frontier: Rethinking Science and Innovation Policy*, edited by Adam B. Jaffe and Benjamin F. Jones, 135–68. Chicago: University of Chicago Press, 2015.

Breznitz, Dan. *Innovation and the State*. New Haven, CT: Yale University Press, 2007.

Breznitz, Dan, and Michael Murphree. *Run of the Red Queen: Government, Innovation, Globalization, and Economic Growth in China*. New Haven, CT: Yale University Press, 2011.

"Briefing by Secretary of the Treasury Steven Mnuchin and Director of the National Economic Council Gary Cohn." *The White House*, April 26, 2017. www.whitehouse.gov/the-press -office/2017/04/26/briefing-secretary-treasury-steven-mnuchin-and-director-national.

Briggs, Vernon M. *Immigration and American Unionism*. Ithaca, NY: Cornell University Press, 2001.

Bring Jobs Home Act of 2011/12. S.3364. 112th Cong., 2011/12. www.congress.gov/bill/112th -congress/senate-bill/3364.

Bring Jobs Home Act of 2013/14, S.2569, 113th Cong., 2013/14. www.congress.gov/bill /112th-congress/senate-bill/2569.

Brooks, Stephen G. *Producing Security: Multinational Corporations, Globalization, and the Changing Calculus of Conflict*. Princeton, NJ: Princeton University Press, 2005.

Brownstein, Ronald. "Back to Basics." *National Journal*, September 9, 2010.

Bruche, Gert. "The Emergence of China and India as New Competitors in MNCs' Innovation Networks." *Competition & Change* 13, no. 3 (2009): 267–88.

Brustein, Joshua. "The Secret Way Silicon Valley Uses the H-1B Program." *Bloomberg*, June 6, 2017. www.bloomberg.com/news/articles/2017-06-06/silicon-valley-s-h-1b-secret.

Buderi, Robert, and Gregory T. Huang. *Guanxi (The Art of Relationships): Microsoft, China, and Bill Gates' Plan to Win the Road Ahead*. London: Random House, 2006.

Bureau of Economic Analysis. "Foreign Direct Investment in the U.S., Majority-Owned Bank and Nonbank U.S. Affiliates, Research and Development Expenditures for 2013." *International Data: Direct Investment and Multinational Enterprises*, 2016. www.bea.gov/iTable /index_MNC.cfm.

——. "U.S. Direct Investment Abroad, All Majority-Owned Foreign Affiliates, Research and Development Expenditures for 2013." *International Data: Direct Investment and Multinational Enterprises*, 2016. www.bea.gov/iTable/index_MNC.cfm.

Bureau of Industry and Security. "Revision and Clarification of Deemed Export Related Regulatory Requirements." *Federal Register*, March 28, 2005.

——. "Revisions and Clarification of Deemed Export Related Regulatory Requirements." *Federal Register*, May 31, 2006.

Burghart, Devin. "Mapping the Tea Party Caucus in the 112th Congress." *Institute for Research and Education on Human Rights*, March 17, 2011. https://irehr.org/issue-areas/tea-party -nationalism/tea-party-news-and-analysis/item/355-mapping-the-tea-party-caucus-in -the-112th-congress.

——. "Special Report: The Status of the Tea Party Movement—Part Two." Institute for Research and Education on Human Rights, January 21, 2014. www.irehr.org/2014/01/21 /status-of-tea-party-by-the-numbers/.

Burghart, Devin, and Leonard Zeskind. "Beyond FAIR: The Decline of the Established Anti-immigration Organizations and the Rise of Tea Party Nativism." Kansas City, MO: Institute for Research and Education on Human Rights, 2012. www.irehr.org/issue-areas /tea-party-nationalism/beyond-fair-report.

Calmes, Jackie, and Edmund Andrews. "Obama Seeks to Curb Foreign Tax Havens." *New York Times*, May 4, 2009. www.nytimes.com/2009/05/05/business/05tax.html?pagewanted =all&_r=1&.

Cameron, Doug, and Julian E. Barnes. "Pentagon Criticizes Contractors' R&D." *Wall Street Journal*, November 21, 2014.

Cameron, Doug, and Alistair Barr. "Google Snubs Robotics Rivals, Pentagon." *Wall Street Journal*, March 5, 2015. www.wsj.com/articles/google-snubs-robotics-rivals-pentagon-1425580734.

Cao, Cong, Ning Li, Xia Li, and Li Liu. "Reforming China's S&T System." *Science* 341, no. 6145 (2013): 460–62.

Carlson, Tucker, and Bill Press. "Debating Immigration Policy." *Crossfire*. CNN, October 24, 2001.

Carroll, Joseph. "American Public Opinion About Immigration." *Gallup.com*, July 26, 2005. www.gallup.com/poll/14785/Immigration.aspx.

Caviedes, Alexander A. *Prying Open Fortress Europe: The Turn to Sectoral Labor Migration*. Lanham, MD: Lexington, 2010.

Center for Immigration Studies. "Are Foreign Students Good for America?" Panel discussion transcript. Rayburn Building, House of Representatives, Washington, DC, June 25, 2002. www.cis.org/sites/cis.org/files/articles/2002/foreignstudents.html.

Center for Responsive Politics. "Interest Groups." *OpenSecrets.org.* Accessed March 16, 2017. www.opensecrets.org/industries/.

——. "Lobbying Spending Database." *OpenSecrets.org.* Accessed March 3, 2017. www.open secrets.org/lobby/.

——. "Lobbying Spending Database—Microsoft." *OpenSecrets.org,* 2015. www.opensecrets.org /lobby/clientsum.php?id=D000000115&year=2013.

Cerna, Lucie. "Attracting High-Skilled Immigrants: Policies in Comparative Perspective." *International Migration* 52, no. 3 (June 2014): 69–84.

——. "The EU Blue Card: Preferences, Policies, and Negotiations between Member States." *Migration Studies* 2, no. 1 (March 1, 2014): 73–96.

——. "The Varieties of High-Skilled Immigration Policies: Coalitions and Policy Outputs in Advanced Industrial Countries." *Journal of European Public Policy* 16, no. 1 (January 2009): 144–61.

Chaddock, Gail Russell. "Eric Cantor Upset Stuns GOP, Revives Tea Party." *Christian Science Monitor,* June 11, 2014. www.csmonitor.com/USA/Elections/2014/0611/Eric-Cantor-upset -stuns-GOP-revives-tea-party.

Chakravorty, Sanjoy, Devesh Kapur, and Nirvikar Singh. *The Other One Percent: Indians in America.* Oxford: Oxford University Press, 2016.

Chase, Kerry A. "Moving Hollywood Abroad: Divided Labor Markets and the New Politics of Trade in Services." *International Organization* 62, no. 4 (2008): 653–87.

Chellaraj, Gnanaraj, Keith E. Maskus, and Aaditya Mattoo. "The Contribution of International Graduate Students to US Innovation." *Review of International Economics* 16, no. 3 (August 2008): 444–62.

Chen, Weihua. "US Immigration Reform a Challenge for China." *China Daily,* February 8, 2013. http://europe.chinadaily.com.cn/opinion/2013-02/08/content_16216633.htm.

Chendakera, Nair. "U.S., India Move to Boost Cooperation in R&D." *Electronic Engineering Times,* March 27, 2000.

Cheng, Ruowei. "Woguo Liushi Dingjian Rencai Shu Ju Shijie Shouwei [China's Loss of Top Talent Is Greatest in the World]." *Renmin Ribao [People's Daily],* June 6, 2013. http:// finance.people.com.cn/n/2013/0606/c1004-21754321.html.

Cheung, Tai Ming, and Bates Gill. "Trade Versus Security: How Countries Balance Technology Transfers with China." *Journal of East Asian Studies* 13, no. 3 (2013): 443–56.

Chin, Hey-Kyung Koh. *Open Doors 2002: Report on International Educational Exchange.* New York: Institute for International Education, 2002.

"China Encourages More VC Funding, Promises Foreign Firms Equal Treatment." *Reuters,* September 1, 2016. www.reuters.com/article/us-china-economy-idUSKCN1175BT.

China Scholarship Council. "2012 Nian Guojia Liuxue Jijin Zizhu Chuguo Liuxue Renyuan Xuanpai Jianzhang [Briefing on the Selection of Awardees for Study Abroad Financial Support in 2012]," November 1, 2011. http://v.csc.edu.cn/Chuguo/739b1b8c11844e5bb211c388563f7da.shtml.

——. "2015 Nian Guojia Gongpai Chuguo Liuxue Xuanpai Jihua Queding [Government Determines Plan for Study Abroad Awards in 2015]." Beijing: Chinese Education Online and Best Choice for Education, October 30, 2014. www.csc.edu.cn/News/2acf973ba1a84ca69f53 86a574771906.shtml.

——. "2017 Nian Guojia Liuxue Jijin Zizhu Chuguo Liuxue Renyuan Xuanpai Jianzhang [Briefing on the Selection of Awardees for Study Abroad Financial Support in 2017]," December 12, 2016. www.csc.edu.cn/article/709.

"China to Overtake US as New Frontier for Global R&D." *People's Daily Online*, January 27, 2014. http://english.peopledaily.com.cn/98649/8523078.html.

Chiswick, Barry R., and Timothy Hatton. "International Migration and the Integration of Labor Markets." In *Globalization in Historical Perspective*, edited by Michael D. Bordo, Alan M. Taylor, and Jeffrey G. Williamson, 65–119. Chicago: University of Chicago Press, 2003.

Chow, Patricia, and Rajika Bhandari. *Open Doors 2010: Report on International Educational Exchange*. New York: Institute for International Education, 2010.

Clayton, Mark. "Academia Becomes Target for New Security Laws." *Christian Science Monitor*, September 24, 2002.

Collier, Andrew K. "Yale Chief Hits at US Student Visa Delays." *South China Morning Post*, November 14, 2003.

Communist Party of China. "Zhonggong Zhongyang Guanyu Jianli Shehuizhuyi Shichang Jingji Tizhi Ruogan Wenti de Jueding [Chinese Communist Party Decision Regarding Several Questions in Building a Socialist Market Economy System]." *Zhongguo Gongchandang Xinwen Wang*, November 14, 1993. http://cpc.people.com.cn/GB/64162/134902/8092314 .html.

"Compete America." *Technology Daily PM*, October 4, 2005.

Comprehensive Immigration Reform Act of 2006. S.2611. 109th Cong., 2006. www.congress.gov /bill/109th-congress/senate-bill/2611.

Confessore, Nicholas. "Borderline Insanity." *Washington Monthly*, May 2002. www.washington monthly.com/features/2001/0205.confessore.html.

Copeland, Dale C. *Economic Interdependence and War*. Princeton, NJ: Princeton University Press, 2014.

——. "Economic Interdependence and War: A Theory of Trade Expectations." *International Security* 20, no. 4 (1996): 5–41.

Costa, Daniel. "Little-Known Temporary Visas for Foreign Tech Workers Depress Wages." *The Hill*, November 11, 2014. http://thehill.com/blogs/pundits-blog/technology/223607-little -known-temporary-visas-for-foreign-tech-workers-depress.

Costlow, Terry. "Senate Set to Vote This Week on Visa-Cap Bill: High Noon Approaches for H-1B Friends, Foes." *Electronic Engineering Times*, October 2, 2000.

Coutin, Susan Bibler. *Nations of Emigrants: Shifting Boundaries of Citizenship in El Salvador and the United States*. Ithaca, NY: Cornell University Press, 2007.

Creating American Jobs and Ending Offshoring Act of 2010. S.3816. 111th Cong., 2010. www .govtrack.us/congress/bills/111/s3816.

Crouzet, François. *The Victorian Economy*. London: Methuen, 1982.

Currie, Chris. "U.S. Public Overwhelmingly Opposed to H-1B Visa Expansion." *IEEE-USA*, September 16, 1998. www.ieeeusa.org/communications/releases/_private/1998/pr091698.html.

Dalrymple, Mary. "Senate Passes Corporate Tax Bill." *Associated Press*, May 12, 2004.

Davis, Christina L., and Sophie Meunier. "Business as Usual? Economic Responses to Political Tensions." *American Journal of Political Science* 55, no. 3 (July 2011): 628–46.

Davis, Todd M. *Open Doors 2000: Report on International Education Exchange.* New York: Institute for International Education, 2000.

"Dealing with Foreign Students and Scholars in an Age of Terrorism: Visa Backlogs and Tracking Systems." *Hearing Before the U.S. House of Representatives Committee on Science.* 108th Cong., March 26, 2003. www.gpo.gov/fdsys/search/home.action.

Deng, Xiaoping. "Gist of Speeches Made in Wuchang, Shenzhen, Zhuhai and Shanghai (From January 18 to February 21, 1992)." *Beijing Review*, February 7, 1994.

Destler, I. M. *American Trade Politics.* New York: Columbia University Press, 2005.

Dickson, Rebecca. "Trump Officials Clamp Down on Worker Visas." *The Hill*, April 6, 2017. http://thehill.com/business-a-lobbying/business-a-lobbying/327507-trump-officials-clamp -down-on-worker-visas.

"Dishiyipi Qianren Jihua Qingnian Rencai, Chuangye Rencai Ruxuan Mingdan [Name List for the Eleventh Batch of Awardees Under the Thousand Youth Talents and Startup Talents Plan]." *Qianren Jihua Wang*, May 13, 2015. http://1000plan.org/qrjh/article/61716.

Dizard, Wilson. "New Law Will Curb Offshoring of Federal IT Work." *Government Computer News*, February 23, 2004.

Docquier, Frédéric, Olivier Lohest, and Abdeslam Marfouk. "Brain Drain in Developing Countries." *The World Bank Economic Review* 21, no. 2 (January 1, 2007): 193–218.

Dossani, Rafiq, and Martin Kenney. "Creating an Environment for Venture Capital in India." *World Development* 30, no. 2 (2002): 227–53.

Drew, Kevin. "Terror Probe Reaches Nation's Campuses." *CNN*, October 25, 2001. http://edition .cnn.com/2001/LAW/10/24/inv.international.students/index.html.

Drezner, Daniel. "State Structure, Technological Leadership and the Maintenance of Hegemony." *Review of International Studies* 27, no. 1 (2001): 3–25.

Dugan, Andrew. "Passing New Immigration Laws Is Important to Americans." *Gallup.com*, July 11, 2013. www.gallup.com/poll/163475/passing-new-immigration-laws-important-americans.aspx.

Dür, Andreas, Patrick Bernhagen, and David Marshall. "Interest Group Success in the European Union: When (and Why) Does Business Lose?" *Comparative Political Studies* 48, no. 8 (2015): 951–83.

Dür, Andreas, and Gemma Mateo. "Public Opinion and Interest Group Influence: How Citizen Groups Derailed the Anti-counterfeiting Trade Agreement." *Journal of European Public Policy* 21, no. 8 (2014): 1199–1217.

Edsall, Thomas B. "Attacks Alter Politics, Shift Focus of Immigration Debate." *Washington Post*, October 15, 2001.

Elmer-DeWitt, Philip. "Apple as the Goose That Laid the Golden Eggs. Five of Them." *Fortune*, November 25, 2013. http://fortune.com/2013/11/25/apple-as-the-goose-that-laid-the -golden-eggs-five-of-them/.

Emery, Gail Repsher. "Industry Fights Dodd Legislation." *Newsbytes News Network*, April 15, 2004.

Englesberg, Paul. "Reversing China's Brain Drain: The Study-Abroad Policy, 1978–1993." In *Great Policies: Strategic Innovations in Asia and the Pacific Basin*, edited by John Dickey Montgomery and Dennis A. Rondinelli, 99–122. Westport, CT: Praeger, 1995.

Ernst and Young. *Globalizing Venture Capital: Global Venture Capital Insights and Trends Report*, 2012. www.ey.com/Publication/vwLUAssets/Globalizing_venture_capital_VC_insights _and_trends_report_CY0227/$FILE/Globalizing%20venture%20capital_VC%20insights %20and%20trends%20report_CY0227.pdf.

Ernst, Dieter. *A New Geography of Knowledge in the Electronics Industry? Asia's Role in Global Innovation Networks*. Honolulu, HI: East-West Center, 2009.

——. *Innovation Offshoring: Exploring Asia's Emerging Role in Global Innovation Networks*. East-West Center Special Report No. 10 (July 2006).

"The Eternal 'Emergency.'" *Wall Street Journal*, August 6, 2010.

European Commission. "The 2016 EU Industrial R&D Investment Scoreboard." *Economics of Industrial Research and Innovation*. Accessed March 13, 2017. http://iri.jrc.ec.europa.eu/score board16.html.

Ewell, Miranda. "Clinton Opposes Higher Visa Cap; Focus on 'Home-Grown' Talent, Commerce Chief Says." *San Jose Mercury News*, January 13, 1998.

FAA Air Transportation Modernization and Safety Improvement Act of 2010. H.R. 1586. 111th Cong. (2010). www.govtrack.us/congress/bills/111/hr1586.

Facchini, Giovanni, and Anna Maria Mayda. "Individual Attitudes Towards Skilled Migration: An Empirical Analysis Across Countries." *The World Economy* 35, no. 2 (2012): 183–96.

Facchini, Giovanni, Anna Maria Mayda, and Prachi Mishra. "Do Interest Groups Affect US Immigration Policy?" *Journal of International Economics* 85, no. 1 (September 2011): 114–28.

Fagerberg, Jan. "Innovation: A Guide to the Literature." In *The Oxford Handbook of Innovation*, edited by Jan Fagerberg, David C. Mowery, and Richard R. Nelson, 1–26. Oxford: Oxford University Press, 2005.

Fagerberg, Jan, David Mowery, and Richard R. Nelson, eds. *The Oxford Handbook of Innovation*. Oxford: Oxford University Press, 2006.

"FAIR Statement on the Pioneer Fund." *PR Newswire*, March 4, 1998.

Farley, Robert. "9/11 Hijackers and Student Visas." *Factcheck.org*, May 10, 2013. www.factcheck. org/2013/05/911-hijackers-and-student-visas/.

Fearon, James D. "Rationalist Explanations for War." *International Organization* 49, no. 3 (1995): 379–414.

Federation for American Immigration Reform. "Immigration Issues: Foreign Students," May 2012. www.fairus.org/issue/foreign-students.

Feldmann, Linda. "Is the Tea Party Running Out of Steam?" *Christian Science Monitor*, April 12, 2014.

Fiegerman, Seth. "Silicon Valley Throws Big Money at Clinton and Virtually Nothing at Trump." *CNN*, August 23, 2016. http://money.cnn.com/2016/08/23/technology/hillary-clinton-tech -fundraisers/.

Fine, Janice, and Daniel J. Tichenor. "An Enduring Dilemma: Immigration and Organized Labor in Western Europe and the United States." In *The Oxford Handbook of the Politics of International Migration*, edited by Marc Rosenblum and Daniel J. Tichenor, 532–72. Oxford: Oxford University Press, 2012.

Finn, Michael G. "Stay Rates of Foreign Doctorate Recipients from U.S. Universities, 2001." Oak Ridge, TN: Oak Ridge Institute for Science and Education, November 2003. http://orise.orau.gov/files/sep/stay-rates-foreign-doctorate-recipients-2001.pdf.

——. "Stay Rates of Foreign Doctorate Recipients from U.S. Universities, 2011." Oak Ridge, TN: Oak Ridge Institute for Science and Education, January 2014. http://orise.orau.gov/science-education/difference/stay-rates-impact.aspx.

Fischer, Karin. "State Department Promises Speedier Visa Review." *Chronicle of Higher Education*, June 12, 2009.

Fitzgerald, David. "Inside the Sending State: The Politics of Mexican Emigration Control." *International Migration Review* 40, no. 2 (2006): 259–93.

Fletcher, Michael. "Bush Immigration Plan Meets GOP Opposition: Lawmakers Resist Temporary-Worker Proposal." *Washington Post*, January 2, 2005.

Foreign Broadcast Information Service. "Foreign Schooling Policy Remains Unchanged." *FBIS-CHI-88-224*, November 14, 1988.

"Fortune 500." *Fortune*, 2015. http://fortune.com/fortune500/.

"Fortune 500: 1955 Full List." *Fortune*, 2015. http://archive.fortune.com/magazines/fortune/fortune500_archive/full/1955/.

Foulon, Mark, and Christopher A. Padilla. "In Pursuit of Security and Prosperity: Technology Controls for a New Era." *Washington Quarterly* 30, no. 2 (2007): 83–90.

"Four Ways to Tackle H-1B Visa Reform." *IEEE Spectrum*, April 19, 2017.

Frankel, Todd, and Tracy Jan. "Trump's New Travel Ban Raises the Same Silicon Valley Objections." *Washington Post*, March 6, 2017. www.washingtonpost.com/news/the-switch/wp/2017/03/06/trumps-new-travel-ban-raises-the-same-silicon-valley-objections/.

Fraser, Steve. "The Hollowing Out of America." *The Nation*, December 3, 2012. www.thenation.com/article/171563/hollowing-out-america#.

Frates, Chris. "Why the Schumer–Smith Immigration Negotiations Broke Down." *Atlantic*, September 20, 2012. www.theatlantic.com/politics/archive/2012/09/why-the-schumer-smith-immigration-negotiations-broke-down/428835/.

Frauenheim, Ed. "Tech Professionals Group Wary of Offshoring." *CNET News*, March 18, 2004.

Freeman, Christopher. *Technology, Policy, and Economic Performance: Lessons from Japan*. London: Pinter, 1987.

Freeman, Gary P. "National Models, Policy Types, and the Politics of Immigration in Liberal Democracies." *West European Politics* 29, no. 2 (March 1, 2006): 227–47.

Freeman, Gary P., and David K. Hill. "Disaggregating Immigration Policy: The Politics of Skilled Labor Recruitment in the US." *Knowledge, Technology & Policy* 19, no. 3 (2006): 7–26.

Friedel, Robert, and Paul B. Israel. *Edison's Electric Light: The Art of Invention*. Baltimore, MD: Johns Hopkins University Press, 2010.

Fu, Xiaolan, and Yundan Gong. "Indigenous and Foreign Innovation Efforts and Drivers of Technological Upgrading: Evidence from China." *World Development* 39, no. 7 (July 2011): 1213–25.

Fu, Xiaolan, and Hongru Xiong. "Open Innovation in China: Policies and Practices." *Journal of Science and Technology Policy in China* 2, no. 3 (2011): 196–218.

Fuller, Douglas B. *Paper Tigers, Hidden Dragons: Firms and the Political Economy of China's Technological Development*. Oxford: Oxford University Press, 2016.

Fung, Brian. "Democrats' Dilemma on High-Skilled Immigration Reform." *National Journal*, March 6, 2013.

Gaillard, Jacques, and Anne-Marie Gaillard. "Introduction: The International Mobility of Brains—Exodus or Circulation?" *Science, Technology & Society* 2, no. 2 (1997): 195–228.

Gallup, Alec, and Frank Newport. *The Gallup Poll: Public Opinion 2004*. Lanham, MD: Rowman & Littlefield, 2006.

Gamboa, Suzanne. "Senate Vote Sidetracks Immigration Compromise." *Associated Press*, April 7, 2006.

Gandhi, Rajiv. "Keep Pace with Technology." *Indian National Congress*, December 19, 1985. www.inc.in/resources/speeches/315-Keep-Pace-with-Technology.

——. "Revamping the Educational System." *Indian National Congress*, August 29, 1985. www.inc.in/resources/speeches/345-Revamping-the-Educational-System.

Gansler, Jacques S., and William Lucyshyn. *Commercial-Off-the-Shelf (COTS): Doing It Right*. College Park: University of Maryland Center for Public Policy and Private Enterprise, 2008. www.dtic.mil/dtic/tr/fulltext/u2/a494143.pdf.

Gaouette, Nicole. "Immigration Bill Ignites a Grass-Roots Fire on the Right." *Los Angeles Times*, June 24, 2007. http://articles.latimes.com/2007/jun/24/nation/na-immig24.

"General Electric Research Lab History." *Edison Tech Center*, 2015. www.edisontechcenter.org/GEresearchLab.html.

George, Alexander, and Andrew Bennett. *Case Studies and Theory Development in the Social Sciences*. Cambridge, MA: MIT Press, 2005.

Gilens, Martin, and Benjamin I. Page. "Testing Theories of American Politics: Elites, Interest Groups, and Average Citizens." *Perspectives on Politics* 12, no. 3 (2014): 564–81.

Gilpin, Robert. *Global Political Economy: Understanding the International Economic Order*. Princeton, NJ: Princeton University Press, 2001.

——. *U.S. Power and the Multinational Corporation: The Political Economy of Foreign Direct Investment*. New York: Basic Books, 1975.

——. *War and Change in World Politics*. Cambridge: Cambridge University Press, 1983.

Glanz, William. "High-Tech Lobbyist Counts Washington Successes." *Knight-Ridder Tribune Business News*, October 14, 2000.

"G.M. Earnings in '55 Go Over Billion Mark." *Chicago Tribune*, February 3, 1956.

Godwin, R. Kenneth. *One Billion Dollars of Influence: The Direct Marketing of Politics*. Chatham, NJ: Chatham House, 1988.

Gohlke, Josh. "Cipel Was No Security Risk, McGreevey Says: Governor Speaks of Former Aide by Name for First Time." *The Record*, September 10, 2004.

Goldfarb, Zachary A., and Karen Tumulty. "IRS Admits Targeting Conservatives for Tax Scrutiny in 2012 Election." *Washington Post*, May 10, 2013. www.washingtonpost.com/business /economy/irs-admits-targeting-conservatives-for-tax-scrutiny-in-2012-election/2013/05/10 /3b6a0ada-b987-11e2-92f3-f291801936b8_story.html.

Goldstein, Judith. *Ideas, Interests, and American Trade Policy*. Ithaca, NY: Cornell University Press, 1993.

"Gonggu Fazhan Zui Guangfan de Aiguo Tongyi Zhanxian Wei Shixian Zhongguo Meng Tigong Guangfan Liliang Zhichi [Consolidate and Develop the Most Wide-Ranging Patriotic United Front to Provide Extensive Supporting Strength to the Chinese Dream]." *Renmin Ribao [People's Daily]*, May 21, 2015. http://paper.people.com.cn/rmrb/html/2015 -05/21/nw.D110000renmrb_20150521_2-01.htm.

Government of India. Emigration Act of 1983. Pub. L. No. 31 (1983). http://moia.gov.in/write readdata/pdf/emig_act.pdf.

——. Passports Act of 1967. Pub. L. No. 15 (1967). http://passportindia.gov.in/AppOnlineProject /pdf/passports_act.pdf.

Government of India Department of Science and Technology. *Science, Technology and Innovation Policy 2013*. New Delhi: Ministry of Science and Technology, 2013. www.dst.gov.in/sti-policy -eng.pdf.

Greenfield, Heather. "Techies 'Disappointed' by Immigration Bill's Demise." *Technology Daily PM*, June 28, 2007.

Greenwalt, Bill. "We Haven't Won Yet on Export Control Reforms." *Breaking Defense*, November 21, 2013. http://breakingdefense.com/2013/11/we-havent-won-yet-on-export-control-reforms/.

Grieco, Joseph M. "Anarchy and the Limits of Cooperation: A Realist Critique of the Newest Liberal Institutionalism." In *Neorealism and Neoliberalism: The Contemporary Debate*, 116–40. New York: Columbia University Press, 1993.

Gross, Andrew D., and Michael S. Schadewald. "Prospects for U.S. Corporate Tax Reform." *The CPA Journal* 82, no. 1 (January 1, 2012).

"Guancha: Zhongguo Gaoduan Rencai Liushi Lu Jugao Buxia [Observation: The Rate of China's High-End Brain Drain Remains High]." *Zhongguo Jiaoyu Bao [China Education Daily]*, January 20, 2014. http://edu.sina.com.cn/a/2014-01-20/1148238838.shtml.

"Guanyu Gongbu Dishier Pi Guojia Qianren Jihua Qingnian Rencai, Chuanye Rencai Ruxuan Renyuan Mingdan de Gonggao [Proclamation Announcing the Twelfth Batch of Awardees for the Thousand Youth Talents and Start-up Talents Plan]." *Qianren Jihua Wang*, March 14, 2016. http://1000plan.org/qrjh/article/61716.

"Guanyu Gongbu Dishisan Pi Guojia Qianren Jihua Qingnian Xiangmu Chuangye Rencai Xiangmu Ruxuan Renyuan Mingdan de Gonggao [Announcement of the Thirteenth Batch of Awardees for the Youth and Startup Talent Programs of the National Thousand Talents Plan]." *Qianren Jihua Wang*, May 11, 2017. http://1000plan.org/qrjh/article/69239.

"Guanyu Gongbu Diyipi Qingnian Qianren Jihua Yinjin Rencai Mingdan de Gonggao [Proclamation Announcing the First Batch of Names for the Thousand Youth Talents Plan to Attract Talent]." *Qianren Jihua Wang*, November 11, 2011. http://1000plan.org/qrjh/article/18053.

Guo, Di, and Kun Jiang. "Venture Capital Investment and the Performance of Entrepreneurial Firms: Evidence from China." *Journal of Corporate Finance* 22 (2013): 375–95.

Guojia Jiaowei Shehui Kexue Yanjiu yu Yishu Jiaoyusi, Guojia Jiaowei Sixiang Zhengzhi Gongzuosi, and Beijing Shiwei Gaodeng Xuexiao Gongzuo Weiyuanhui. *Wushitian de Huigu Yu Fansi [Looking Back and Reflecting on Fifty Days]*. Beijing: Gaodeng Jiaoyu Chubanshe, 1989.

"Guruduth Banavar." *LinkedIn*, accessed February 11, 2016. www.linkedin.com/in/banavar.

Haddal, Chad C. *Foreign Students in the United States: Policies and Legislation*. Washington, DC: Congressional Research Service, 2008.

Haggard, Stephan. "The Institutional Foundations of Hegemony: Explaining the Reciprocal Trade Agreements Act of 1934." *International Organization* 42, no. 1 (1988): 91–119.

Hahm, Jung S. "American Competitiveness and Workforce Improvement Act of 1998: Balancing Economic and Labor Interests Under the New H-1B Visa Program." *Cornell Law Review* 85 (1999): 1673–1701.

Hainmueller, Jens, and Michael J. Hiscox. "Attitudes Toward Highly Skilled and Low-Skilled Immigration: Evidence from a Survey Experiment." *American Political Science Review* 104, no. 1 (February 2010): 61.

Hannas, William C., James Mulvenon, and Anna B. Puglisi. *Chinese Industrial Espionage: Technology Acquisition and Military Modernization*. New York: Routledge, 2013.

Hart, David. "High-Tech Learns to Play the Washington Game, or the Political Education of Bill Gates and Other Nerds." In *Interest Group Politics*, edited by Allan J. Cigler and Burdett Loomis. 6th ed. Washington, DC: CQ, 2002.

——. "New Economy, Old Politics: The Evolving Role of the High-Technology Industry in Washington, D.C." In *Governance amid Bigger, Better Markets*, edited by Joseph S. Nye and John D. Donahue, 235–65. Washington, DC: Brookings Institution, 2004.

——. "Red, White, and 'Big Blue': IBM and the Business–Government Interface in the United States, 1956–2000." *Enterprise and Society* 8, no. 1 (2007): 1–34.

Hart, David M. "'Business' Is Not an Interest Group: On the Study of Companies in American National Politics." *Annual Review of Political Science* 7 (2004): 47–69.

——. "Political Representation in Concentrated Industries: Revisiting the 'Olsonian Hypothesis.'" *Business and Politics* 5, no. 3 (2003): 261–86.

——. "Understanding Immigration in a National Systems of Innovation Framework." *Science & Public Policy* 34, no. 1 (2007): 45–53.

Hattori, James. "Intel Plans for Future." CNN, September 14, 2002.

Hawthorne, Lesleyanne. "The Growing Global Demand for Students as Skilled Migrants." Washington, DC: Migration Policy Institute, 2008. www.migrationpolicy.org/sites/default/files/publications/intlstudents_0.pdf.

Heath, Jena. "Congressman Switches Focus to High-Tech: Republican, New to Area, Softens Immigration Stance, Trumpets Tech." *Austin American-Statesman*, July 7, 2002.

Helpman, Elhanan. *The Mystery of Economic Growth*. Cambridge, MA: Harvard University Press, 2004.

Hempel, Jessi. "DOD Head Ashton Carter Enlists Silicon Valley to Transform the Military." *Wired*, November 18, 2015. www.wired.com/2015/11/secretary-of-defense-ashton-carter/.

Henderson, Richard M., and Kim B. Clark. "Architectural Innovation: The Reconfiguration of Existing Product Technologies and the Failure of Established Firms." *Administrative Science Quarterly* 35, no. 1 (March 1990): 9–30.

Heppenheimer, T. A. *Turbulent Skies: The History of Commercial Aviation*. New York: Wiley, 1995.

Hicks, Raymond, Helen V. Milner, and Dustin Tingley. "Trade Policy, Economic Interests, and Party Politics in a Developing Country: The Political Economy of CAFTA-DR." *International Studies Quarterly* 58, no. 1 (March 2014): 106–17.

Hiebert, Murray. "United States: The Cost of Security." *The Far Eastern Economic Review*, November 28, 2002.

Higgins, John K. "Proposed 2016 Federal Budget Plumps IT Spending by $2B." *E-Commerce Times*, March 11, 2015. www.ecommercetimes.com/story/81805.html.

High Level Committee on the Indian Diaspora. "The Indian Diaspora." New Delhi: Ministry of External Affairs, 2001. http://indiandiaspora.nic.in/contents.htm.

Hira, Ron. "Implications of Offshore Outsourcing." Paper presented at the Globalization, Employment, and Economic Development Workshop. Sloan Workshop Series in Industry Studies, Rockport, MA, January 3, 2004.

Hira, Ronil. Testimony. *Hearing Before the Senate Judiciary Committee on the Border Security, Economic Opportunity, and Immigration Modernization Act, S.744*. 113th Cong. (April 22, 2013).

——. "New Data Show How Firms Like Infosys and Tata Abuse the H-1B Program." *Economic Policy Institute*, February 19, 2015. www.epi.org/blog/new-data-infosys-tata-abuse-h-1b-program/.

——. "U.S. Immigration Regulations and India's Information Technology Industry." *Technological Forecasting and Social Change* 71, no. 8 (October 2004): 837–54.

"Hispanic Business Magazine Includes Twelve NCLR Affiliates Among Its Top Twenty-Five Nonprofits." *Targeted News Service*, June 9, 2011.

Hollifield, James F. "The Emerging Migration State." *International Migration Review* 38, no. 3 (2004): 885–912.

Holyoke, Thomas T. "The Interest Group Effect on Citizen Contact with Congress." *Party Politics* 19, no. 6 (November 1, 2013): 925–44.

Hook, Janet. "Tea Party Faces Test of Its Clout in Primaries." *Wall Street Journal*, February 25, 2014.

Horowitz, Daniel. "Gang Immigration Bill (S.744) Is Comprehensively Flawed." *Madison Project*. May 1, 2013. http://madisonproject.com/2013/05/gang-immigration-bill-s-744-is-comprehensively-flawed/.

Horowitz, Jason. "Marco Rubio Pushed for Immigration Reform with Conservative Media." *New York Times*, February 27, 2016. www.nytimes.com/2016/02/28/us/politics/marco-rubio-pushed-for-immigration-reform-with-conservative-media.html.

Horowitz, Michael C. *The Diffusion of Military Power: Causes and Consequences for International Politics*. Princeton, NJ: Princeton University Press, 2010.

Hounshell, David A., and John Kenly Smith. *Science and Corporate Strategy: Du Pont R&D, 1902–1980*. Cambridge: Cambridge University Press, 1988.

"House Dems, AFL-CIO Discuss H-1B Visas." *Congress Daily*, March 14, 2000.

Hufbauer, Gary Clyde, Theodore H. Moran, and Lindsay Oldenski. *Outward Foreign Direct Investment and US Exports, Jobs, and R&D: Implications for US Policy*. Washington, DC: Peterson Institute for International Economics, 2013.

Hughes, Llewelyn. *Globalizing Oil: Firms and Oil Market Governance in France, Japan, and the United States*. New York: Cambridge University Press, 2014.

Humphery-Jenner, Mark, and Jo-Ann Suchard. "Foreign Venture Capitalists and the Internationalization of Entrepreneurial Companies: Evidence from China." *Journal of International Business Studies* 44, no. 6 (2013): 607–21.

IBM. "IBM Research: Global Labs." *IBM*, accessed April 3, 2015. www.research.ibm.com/labs/.

"IEEE Advocates Limits on Offshore Outsourcing." *Information Week*, March 15, 2004. www.informationweek.com/ieee-advocates-limits-on-offshore-outsourcing/d/d-id/1023812.

Ikenberry, G. John. *After Victory: Institutions, Strategic Restraint, and the Rebuilding of Order after Major Wars*. Princeton, NJ: Princeton University Press, 2001.

Ikenberry, G. John, David A. Lake, and Michael Mastanduno. "Introduction: Approaches to Explaining American Foreign Economic Policy." In *The State and American Foreign Economic Policy*, edited by G. John Ikenberry and David A. Lake. Ithaca, NY: Cornell University Press, 1988.

"India May Drag US to WTO for Hiking H-1B Visa Fee." *Times of India*, August 17, 2010. http://timesofindia.indiatimes.com/business/india-business/India-may-drag-US-to-WTO-for-hiking-H-1B-visa-fee/articleshow/6325497.cms.

Information Technology Industry Council. "ITI on Trump Tax Reform Principles," *Information Technology Industry Council*, April 26, 2017. www.itic.org/news-events/news-releases/iti-on-trump-tax-reform-principles.

Institute for International Education. "International Student Totals by Place of Origin, 2013/14–2014/15." *Open Doors Report on International Educational Exchange*, 2015. www.iie.org/Research-and-Publications/Open-Doors/Data/International-Students/All-Places-of-Origin/2013-15.

——. "Open Doors Data: Fast Facts." *Institute for International Education*, 2014. www.iie.org/Research-and-Publications/Open-Doors/Data/Fast-Facts.

——. "Top Twenty-Five Places of Origin of International Students, 2014/15 and 2015/16." *Open Doors Report on International Educational Exchange*. New York: Institute for International Education, 2016. www.iie.org/opendoors.

Institute for Regional Studies. "Silicon Valley Index 2017." *Silicon Valley Indicators*, February 2017. http://jointventure.org/images/stories/pdf/index2017.pdf.

"Intel China Research Center." *Intel*. Accessed May 6, 2014. www.intel.com/cd/corporate/icrc/apac/eng/about/167066.htm.

Jackson, James K. *The Committee on Foreign Investment in the United States (CFIUS)*. Washington, DC: Congressional Research Service, 2014.

Jacobsen, Annie. *Operation Paperclip: The Secret Intelligence Program That Brought Nazi Scientists to America*. New York: Little, Brown, 2014.

Janardhanan, Arun. "US Move to Lure Science Grads Worries India." *Times of India—Chennai Edition*, February 4, 2013.

Jaruzelski, Barry, Volker Staack, and Aritomo Shinozaki. "2016 Global Innovation 1000 Study." *PwC.* Accessed March 15, 2017. www.strategyand.pwc.com/innovation1000.

——. "Software-as-a-Catalyst." *Strategy+Business*, October 25, 2016. www.strategy-business.com /feature/Software-as-a-Catalyst?gko=7a1ae.

Jensen, J. Bradford. *Global Trade in Services: Fear, Facts, and Offshoring.* Washington, DC: Peterson Institute for International Economics, 2011.

Jeronimides, H. Rosemary. "The H-1B Visa Category: A Tug of War." *Georgetown Immigration Law Journal* 7 (1993): 367–91.

Jischke, Martin. "Addressing the New Reality of Current Visa Policy on International Students and Researchers." Testimony. *Hearing Before the U.S. Senate Committee on Foreign Relations.* 108th Cong. (October 6, 2004). www.gpo.gov/fdsys/search/home.action.

Jobs and Trade Network. "Jobs and Trade Network to Hold Press Luncheon with U.S. Sen. Dodd: National Fair Trade Group to Advocate Against Outsourcing Policies," February 20, 2004. http://epi.3cdn.net/664ff3156937cef731_00m6bxahk.pdf.

John, Sujit. "Cisco Needs to Align with Indian Government's Goals." *Times of India*, July 2, 2014.

John, Sujit, and Shilpa Phadnis. "For MNCs, India Still an R&D Hub and It's Growing." *Times of India*, March 2, 2017. http://timesofindia.indiatimes.com/city/bengaluru/for-mncs-india -still-an-rd-hub-and-its-growing/articleshow/57421665.cms.

Johnson, Marlene. Testimony. "Addressing the New Reality of Current Visa Policy on International Students and Researchers." *Hearing Before the U.S. Senate Committee on Foreign Relations.* 108th Cong. (October 6, 2004). www.gpo.gov/fdsys/search/home.action.

Kaar, Robbert van het, and Marianne Grünell. "Industrial Relations in the Information and Communications Technology Sector." *Eurofound*, August 27, 2001. www.eurofound .europa.eu/observatories/eurwork/comparative-information/industrial-relations-in-the -information-and-communications-technology-sector.

Kamieniecki, Sheldon. *Corporate America and Environmental Policy: How Often Does Business Get Its Way?* Stanford, CA: Stanford University Press, 2006.

Kaplan, Rebecca. "How the Tea Party Came Around on Immigration." *National Journal*, March 21, 2013.

Kapur, Devesh. *Diaspora, Development, and Democracy: The Domestic Impact of International Migration from India.* Princeton, NJ: Princeton University Press, 2010.

——. "Indian Higher Education." In *American Universities in a Global Market*, edited by Charles T. Clotfelter, 305–34. Chicago: University of Chicago Press, 2010.

Kelly, Kate, Rachel Abrams, and Alan Rappeport. "Trump Is Said to Abandon Contentious Border Tax on Imports." *New York Times*, April 25, 2017. www.nytimes.com/2017/04/25/us /politics/orrin-hatch-trump-tax-cuts-deficit-economy.html.

Kennedy, Andrew B. "China's Search for Renewable Energy: Pragmatic Techno-Nationalism." *Asian Survey* 53, no. 5 (2013): 909–30.

——. "India's Nuclear Odyssey: Implicit Umbrellas, Diplomatic Disappointments, and the Bomb." *International Security* 36, no. 2 (2011): 120–53.

——. *The International Ambitions of Mao and Nehru: National Efficacy Beliefs and the Making of Foreign Policy.* New York: Cambridge University Press, 2012.

——. "Powerhouses or Pretenders? Debating China's and India's Emergence as Technological Powers." *The Pacific Review* 28, no. 2 (2015): 281–302.

——. "Slouching Tiger, Roaring Dragon: Comparing India and China as Late Innovators." *Review of International Political Economy* 23, no. 2 (2016): 1–28.

——. "Unequal Partners: U.S. Collaboration with China and India in Research and Development." *Political Science Quarterly* 132, no. 1 (2017).

Kennedy, Paul. *The Rise and Fall of the Great Powers.* New York: Vintage, 1987.

Keohane, Robert O. "The Old IPE and the New." *Review of International Political Economy* 16, no. 1 (2009): 34–46.

"Kerry Tax Plan Proposes to Slow Loss of US Jobs Overseas." *Dow Jones International News*, March 26, 2004.

Kessler, Glen, and Kevin Sullivan. "Powell Cautious About Immigration Changes." *Washington Post*, November 10, 2004.

Khadria, Binod. "Brain Drain, Brain Gain, India." In *The Encyclopedia of Global Human Migration*, edited by Immanuel Ness, 743–49. New York: Wiley-Blackwell, 2013.

Khimm, Suzy. "Why a Rare Bipartisan Consensus on Immigration Totally Fell Apart." *Washington Post*, September 21, 2012. www.washingtonpost.com/blogs/ezra-klein/wp/2012/09/21/why-a -rare-bipartisan-consensus-on-immigration-totally-fell-apart/.

Kiely, Kathy. "Immigration Overhaul Crumbles in Senate Vote." *USA Today*, June 29, 2007.

Kim, Dong Jung. "Cutting Off Your Nose? A Reigning Power's Commercial Containment of a Military Challenger." Ph.D. diss., University of Chicago, 2015.

——. "Trading with the Enemy? The Futility of US Commercial Countermeasures Against the Chinese Challenge." *Pacific Review*, November 2, 2016, 1–20.

Kim, In Song. "Political Cleavages Within Industry: Firm Level Lobbying for Trade Liberalization." *American Political Science Review* 111 (2017): 1–20.

Kim, Linsu. *Imitation to Innovation: The Dynamics of Korea's Technological Learning.* Cambridge, MA: Harvard Business Press, 1997.

Kim, Sung-Young. "Transitioning from Fast-Follower to Innovator: The Institutional Foundations of the Korean Telecommunications Sector." *Review of International Political Economy* 19, no. 1 (February 2012): 140–68.

Kimball, David C., Frank R. Baumgartner, Jeffrey M. Berry, Marie Hojnacki, Beth L. Leech, and Bryce Summary. "Who Cares About the Lobbying Agenda?" *Interest Groups & Advocacy* 1, no. 1 (May 2012): 5–25.

Kindleberger, Charles Poor. *The World in Depression, 1929–1939.* Berkeley: University of California Press, 1973.

King, Neil, and Elizabeth Williamson. "Business Fends Off Tax Hit: Obama Administration Shelves Plan to Change How U.S. Treats Overseas Profits." *Wall Street Journal*, October 13, 2009.

Kocieniski, David. "For Trenton's Lame Duck, the Question Is, 'How Lame?'" *New York Times*, September 12, 2004.

Koehn, Peter H. "Developments in Transnational Research Linkages: Evidence from US Higher-Education Activity." *Journal of New Approaches in Educational Research* 3, no. 2 (2014): 52–58.

Koffler, Keith. "Business Coalition Rewrites Lexicon for Jobs 'Outsourcing.'" *Congress Daily*, March 2, 2004.

Kollman, Ken. *Outside Lobbying: Public Opinion and Interest Group Strategies*. Princeton, NJ: Princeton University Press, 1998.

KPMG and CB Insights. *Venture Pulse 2016*. New York: CB Insights, April 13, 2016. https://www .cbinsights.com/research-venture-capital-Q1-2016.

Krasner, Stephen D. "State Power and the Structure of International Trade." *World Politics* 28, no. 3 (1976): 317–47.

Krause, Reinhardt. "Tech Firms Pushing for More H-1B Visas for Skilled Workers." *Investor's Business Daily*, April 11, 2007.

Kudrle, Robert T., and Davis B. Bobrow. "U.S. Policy Toward Foreign Direct Investment." *World Politics* 34, no. 3 (April 1982): 353–79.

Kumar, Nirmalya, and Phanish Puranam. *India Inside: The Emerging Innovation Challenge to the West*. Cambridge, MA: Harvard Business Press, 2012.

Kuptsch, Christiane, and Eng Fong Pang. "Introduction." In *Competing for Global Talent*, edited by Christiane Kuptsch and Eng Fong Pang, 1–8. Geneva: International Labor Office, 2006.

Lake, David A. "Leadership, Hegemony, and the International Economy: Naked Emperor or Tattered Monarch with Potential?" *International Studies Quarterly* 37, no. 4 (December 1993): 459.

——. "Open Economy Politics: A Critical Review." *The Review of International Organizations* 4, no. 3 (September 2009): 219–44.

Lampton, David M. *A Relationship Restored: Trends in U.S.–China Educational Exchanges, 1978–1984*. Washington, DC: National Academies Press, 1986.

Landes, David S. *The Unbound Prometheus: Technological Change and Industrial Development in Western Europe from 1750 to the Present*. Cambridge: Cambridge University Press, 2003.

Lane, Jason, and Kevin Kinser. "Is Today's University the New Multinational Corporation?" *The Conversation*, June 5, 2015. http://theconversation.com/is-todays-university-the-new -multinational-corporation-40681.

Lawrence, Jill. "The Myth of Marco Rubio's Immigration Problem." *National Journal*, July 15, 2013.

Lazonick, William. *Sustainable Prosperity in the New Economy? Business Organization and High-Tech Employment in the United States*. Kalamazoo, MI: W. E. Upjohn Institute, 2009.

Leibovich, Mark. "High Tech Is King of the Hill." *Washington Post*, October 16, 1998.

Lewin, Tamar. "Foreign Students Bring Cash, and Changes: U.S. Colleges Welcome Funds, but Some In-State Applicants Feel Left Out." *International Herald Tribune*, February 6, 2012.

Liberto, Jennifer, and Dana Bash. "Bill to Hike Taxes on Overseas Jobs Fails Senate Test Vote." *CNN*, September 28, 2010. http://money.cnn.com/2010/09/28/news/economy/Outsource _jobs_bill_dead/.

Lieberman, Joseph I. *Offshore Outsourcing and America's Competitive Edge: Losing Out in the High Technology R&D and Services Sectors*. Washington, DC: Office of Senator Joseph I. Lieberman, May 11, 2004.

Lipton, Eric, and Somini Sengupta. "Latest Product of Tech Firms: Immigrant Bill." *New York Times*, May 5, 2013.

Lissardy, Gerardo. "Leading Hispanic Group Joins Immigrant-Rights Coalition." *EFE News Service*, May 12, 2006.

Liu, Charlotte, Nick Campbell, Ed Gerstner, Amy Lin, Piao Li, Stephen Pincock, Chandler Gibbons, et al. "Turning Point: Chinese Science in Transition." Shanghai: Nature Publishing Group, November 2015. www.nature.com/press_releases/turning_point.pdf.

Liu, Hong, and Els van Dongen. "China's Diaspora Policies as a New Mode of Transnational Governance." *Journal of Contemporary China* 25, no. 102 (2016): 1–17.

Lizza, Ryan. "Getting to Maybe." *New Yorker*, June 24, 2013. www.newyorker.com/magazine/2013/06/24/getting-to-maybe.

Lochhead, Carolyn. "Bill to Boost Tech Visas Sails Through Congress." *San Francisco Chronicle*, October 4, 2000. www.sfgate.com/news/article/Bill-to-Boost-Tech-Visas-Sails-Through-Congress-2735682.php.

——. "Visa Plan Angers Silicon Valley." *SFGate*, June 7, 2007. www.sfgate.com/politics/article/VISA-PLAN-ANGERS-SILICON-VALLEY-Immigration-2588829.php.

Lohr, Steve. "Many New Causes for Old Problem of Jobs Lost Abroad." *New York Times*, February 15, 2004.

——. "New Economy; Offshore Jobs in Technology: Opportunity or a Threat?" *New York Times*, December 22, 2003. www.nytimes.com/2003/12/22/business/new-economy-offshore-jobs-in-technology-opportunity-or-a-threat.html.

Lopez, Mark Hugo. "The Hispanic Vote in the 2008 Election." *Pew Research Center*, November 5, 2008. www.pewhispanic.org/2008/11/05/the-hispanic-vote-in-the-2008-election/.

"Lott Wants Agreement with Dems on H-1B Visa Measure." *Congress Daily*, September 14, 2000.

Lucas, Louise. "US Concerns Grow Over Chinese Chip Expansion." *Financial Times*, January 16, 2017. www.ft.com/content/fb2e4454-c36e-11e6-9bca-2b93a6856354.

Ludden, Jennifer. "Strange Bedfellows Join Forces for Immigration Reform." *National Public Radio: All Things Considered*, January 19, 2006.

Lundvall, Bengt-Åke. *National Systems of Innovation: Toward a Theory of Innovation and Interactive Learning*. London: Pinter, 1992.

Lynch, David J. "Does Tax Code Encourage U.S. Companies to Cut Jobs at Home? Presidential Candidates Target Corporate Tax Breaks for Offshoring." *USA Today*, March 21, 2008.

Ma, Chi. "Famous Science Projects Face Axe in Funding Overhaul." *China Daily*, January 8, 2015. www.chinadaily.com.cn/china/2015-01/08/content_19275863.htm.

Madden, Mike. "Millions Spent Lobbying on Immigration in Last Congress." *Gannett News Service*, January 26, 2007.

Malhotra, Neil, Yotam Margalit, and Cecilia Hyunjung Mo. "Economic Explanations for Opposition to Immigration: Distinguishing Between Prevalence and Conditional Impact." *American Journal of Political Science* 57, no. 2 (April 2013): 391–410.

Mansfield, Edward D., and Diana C. Mutz. "US Versus Them: Mass Attitudes Toward Offshore Outsourcing." *World Politics* 65, no. 4 (2013): 571–608.

Mansfield, Edward D., and Brian Pollins. "Interdependence and Conflict: An Introduction." In *Economic Interdependence and International Conflict: New Perspectives on an Enduring Debate*, edited by Edward D. Mansfield and Brian Pollins. Ann Arbor: University of Michigan Press, 2003.

Marin, Anabel, and Subash Sasidharan. "Heterogeneous MNC Subsidiaries and Technological Spillovers: Explaining Positive and Negative Effects in India." *Research Policy* 39, no. 9 (November 2010): 1227–41.

Marro, Nick. "Foreign Company R&D: In China, for China." *China Business Review*, June 1, 2015.

Marsan, Carolyn Duffy. "A Political Hot Potato: Legislatures Juggle Offshore Outsourcing Regulations." *Network World*, July 5, 2004.

Marschall, Daniel, and Laura Clawson. *Sending Jobs Overseas: The Cost to America's Economy and Working Families*. Washington, DC: Working America and the AFL-CIO, 2010.

Mastanduno, Michael. *Economic Containment: CoCom and the Politics of East–West Trade*. Ithaca, NY: Cornell University Press, 1992.

Matloff, Norman. "On the Need for Reform of the H-1B Non-immigrant Work Visa in Computer-Related Occupations." *University of Michigan Journal of Law Reform* 36, no. 4 (Fall 2003): 815–914.

McConnell, Mitch. (KY). "Immigration." *Congressional Record*, 110th Cong., 153, no. 106 (June 28, 2007): S8674.

"McGreevey Edict Restricts Outsourcing by Agencies." *Congress Daily*, September 13, 2004.

McGreevey, James E. *The Confession*. New York: Harper Collins, 2006.

McKinnon, John D. "Plan Would Raise Taxes on Businesses." *Wall Street Journal*, February 2, 2010. www.wsj.com/articles/SB10001424052748704107204575039073372259004.

Mearsheimer, John J. *The Tragedy of Great Power Politics*. New York: W. W. Norton, 2001.

Meckler, Laura. "House Immigration Bills Are Still in the Mix." *Wall Street Journal*, April 18, 2014. www.wsj.com/articles/SB10001424052702304626304579508091839546088.

——. "Immigration-Bill Pressure Backfires; Overhaul Backers Target Majority Whip, but Tactic Provokes Response from Opponents." *Wall Street Journal*, December 25, 2013. www.wsj.com/articles/SB10001424052702304244904579276403694719232.

——. "Visas Could Aid Graduates." *Wall Street Journal*, October 22, 2011.

Meijer, Hugo. *Trading with the Enemy: The Making of US Export Control Policy Toward the People's Republic of China*. Oxford: Oxford University Press, 2016.

Menz, Georg. *The Political Economy of Managed Migration: Nonstate Actors, Europeanization, and the Politics of Designing Migration Policies*. Oxford: Oxford University Press, 2008.

Meyers, Jessica. "Tech Companies See Few Big Gains in Obama's Executive Action." *Boston Globe*, November 24, 2014. www.bostonglobe.com/news/nation/2014/11/24/tech-companies-see-few-big-gains-obama-executive-action/dauDJujkOhe1qx5ZQTScoM/story.html.

Milbank, Dana. "Jabs and All, the Ides of March Arrives Late." *Washington Post*, June 29, 2007.

Miller, Matthew. "Spy Scandal Weighs on U.S. Tech Firms in China, Cisco Takes Hit." *Reuters*. November 14, 2013. www.reuters.com/article/2013/11/14/us-china-cisco-idUSBRE9AD0J420131114.

Milner, Helen V. *Resisting Protectionism: Global Industries and the Politics of International Trade.* Princeton, NJ: Princeton University Press, 1988.

Milner, Helen, and David B. Yoffie. "Between Free Trade and Protectionism: Strategic Trade Policy and a Theory of Corporate Trade Demands." *International Organization* 43, no. 2 (Spring 1989): 239–72.

Milton, Laurie P. "An Identity Perspective on the Propensity of High-Tech Talent to Unionize." *Journal of Labor Research* 24, no. 1 (2003): 31–53.

Mitchell, Brian. "Frist's Border Control-Only Bill Spurs Broad Immigration Deals." *Investor's Business Daily*, March 20, 2006.

Modelski, George, and William R. Thompson. *Leading Sectors and World Powers: The Coevolution of Global Politics and Economics.* Columbia, SC: University of South Carolina Press, 1996.

Modi, Narendra. "Modi Speaks in San Jose: The Indian Prime Minister in His Own Words." *SiliconValleyOneWorld*, September 27, 2015. www.siliconvalleyoneworld.com/2015/09/30/modi-speaks-in-san-jose-the-indian-prime-minister-in-his-own-words/.

——. "PM's Address to the Nation from the Ramparts of the Red Fort on the Sixty-Eighth Independence Day," August 15, 2014. http://pmindia.gov.in/en/news_updates/text-of-pms-address-in-hindi-to-the-nation-from-the-ramparts-of-the-red-fort-on-the-68th-independence-day/.

——. "PM's Speech to 104th Session of the Indian Science Congress, Tirupati (Full Text)." *Microfinance Monitor*, January 3, 2017. www.microfinancemonitor.com/pms-speech-to-104th-session-of-the-indian-science-congress-tirupati-full-text/43799.

——. "Text of PM Shri Narendra Modi's Address at the 102nd Indian Science Congress." *Official Website of Narendra Modi*, January 3, 2015. www.narendramodi.in/text-of-pm-shri-narendra-modis-address-at-the-102nd-indian-science-congress.

Moe, Espen. *Governance, Growth and Global Leadership: The Role of the State in Technological Progress, 1750–2000.* Hampshire, UK: Ashgate, 2013.

——. "Mancur Olson and Structural Economic Change: Vested Interests and the Industrial Rise and Fall of the Great Powers." *Review of International Political Economy* 16, no. 2 (June 26, 2009): 202–30.

Mokyr, Joel. *The Lever of Riches: Technological Creativity and Economic Progress.* New York: Oxford University Press, 1990.

Moravcsik, Andrew. "Liberal Theories of International Law." In *Interdisciplinary Perspectives on International Law and International Relations: The State of the Art*, edited by Jeffrey L. Dunoff and Mark A. Pollack, 83–118. Cambridge: Cambridge University Press, 2013.

——. "Taking Preferences Seriously: A Liberal Theory of International Politics." *International Organization* 51, no. 4 (1997): 513–53.

Morrison, Bruce. Testimony. *Hearing Before the Committee on the Judiciary of the United States House of Representatives Subcommittee on Immigration Policy and Enforcement.* 113th Cong. (March 5, 2013).

Morrison, James Ashley. "Before Hegemony: Adam Smith, American Independence, and the Origins of the First Era of Globalization." *International Organization* 66, no. 3 (2012): 395–428.

Moscoso, Eunice. "Once Gung-Ho, Businesses See Flaws in Immigration Bill: Tech Sector Particularly Disturbed by Potential Changes in Visa Program." *The Atlanta Journal-Constitution*, June 3, 2007.

Moser, Petra, Alessandra Voena, and Fabian Waldinger. "German Jewish Émigrés and US Invention." *The American Economic Review* 104, no. 10 (2014): 3222–55.

Mowery, David C., and Nathan Rosenberg. "The U.S. National Innovation System." In *National Innovation Systems: A Comparative Analysis*, edited by Richard R. Nelson, 29–75. Oxford: Oxford University Press, 1993.

Mowery, David. C., and Bhaven N. Sampat, "Universities in National Innovation Systems." In *The Oxford Handbook of Innovation*, edited by Jan Fagerberg, David Mowery, and Richard R. Nelson. Oxford: Oxford University Press, 2006.

Mrinalini, N., Pradosh Nath, and G. D. Sandhya. "Foreign Direct Investment in R&D in India." *Current Science* 105, no. 6 (September 2013): 767–73.

Mueller, Dennis C. "First-Mover Advantages and Path Dependence." *International Journal of Industrial Organization* 15, no. 6 (October 1997): 827–50.

Mufson, Steven. "Once a Recession Remedy, GM's Empire Falls." *Washington Post*, June 2, 2009.

Mukem, Anne C. "Firebrand Tancredo Puts Policy Over Party Line." *Denver Post*, November 27, 2005.

Munoz, Sara Schaefer. "Firms Push to Expand Visa Program." *Wall Street Journal*, March 11, 2004.

Munro, Neil. "IT Industry, Hispanics Team Up On Immigration." *National Journal*, April 9, 2010.

Murmann, Johann Peter, and Ralph Landau. "On the Making of Competitive Advantage: The Development of the Chemical Industry in Britain and Germany Since 1850." In *Chemicals and Long-Term Economic Growth: Insights from the Chemical Industry*, edited by Ashish Arora, Ralph Landau, and Nathan Rosenberg, 27–70. New York: Wiley, 1998.

Murphy, Caryle, and Nurith C. Aizenman. "Foreign Students Navigate Labyrinth of New Laws: Slip-Ups Overlooked Before 9/11 Now Grounds for Deportation." *Washington Post*, June 9, 2003.

Murphy, Colum, and Lilian Lin. "For China's Jobseekers, Multinational Companies Lose Their Magic." *Wall Street Journal*, April 3, 2014. http://blogs.wsj.com/chinarealtime/2014/04/03/for-chinas-jobseekers-multinational-companies-lose-their-magic/?mod=chinablog.

Murray, Michael A. "Defining the Higher Education Lobby." *The Journal of Higher Education* 47, no. 1 (January 1976): 79–92.

Murray, Shailagh. "Careful Strategy Is Used to Derail Immigration Bill." *Washington Post*, June 8, 2007.

——. "In Senate and on Trail, Democrats Target Jobs Moving Abroad." *Washington Post*, June 9, 2010.

Nakamura, David. "Conservatives Split on Immigration Bill's Price Tag." *Washington Post*, May 7, 2013.

Narula, Rajneesh, and Antonello Zanfei. "Globalization of Innovation: The Role of Multinational Enterprises." In *The Oxford Handbook of Innovation*, edited by Jan Fagerberg, David C. Mowery, and Richard R. Nelson, 318–45. Oxford: Oxford University Press, 2005.

National Council of La Raza. *2011 Annual Report*. Washington, DC: National Council of La Raza, 2011. http://publications.nclr.org/handle/123456789/2.

National Foundation for American Policy. *NFAP Policy Brief: Anti-outsourcing Efforts Down but Not Out*. Arlington, VA: National Foundation for American Policy, April 2007. www.nfap .com/pdf/0407OutsourcingBrief.pdf.

National Science Foundation. *Science and Engineering Indicators 2004*. Arlington, VA: National Science Foundation, 2004.

——. *Science and Engineering Indicators 2012*. Arlington, VA: National Science Foundation, 2012.

——. *Science and Engineering Indicators 2014*. Arlington, VA: National Science Foundation, 2014.

——. *Science and Engineering Indicators 2016*. Arlington, VA: National Science Foundation, 2016.

——. "Table 53: Doctorate Recipients with Temporary Visas Intending to Stay in the United States After Doctorate Receipt, by Country of Citizenship: 2007–13." *Science and Engineering Doctorates*, December 2014. www.nsf.gov/statistics/sed/2013/data-tables.cfm.

Naujoks, Daniel. *Migration, Citizenship, and Development: Diasporic Membership Policies and Overseas Indians in the United States*. New Delhi: Oxford University Press, 2013.

"Navigating China's Tech Jungle." *Business Times*, September 1, 2012.

Nelson, David, and Susan Webb Yackee. "Lobbying Coalitions and Government Policy Change: An Analysis of Federal Agency Rulemaking." *The Journal of Politics* 74, no. 2 (2012): 339–53.

Nelson, Richard R., ed. *National Innovation Systems: A Comparative Analysis*. Oxford: Oxford University Press, 1993.

"New Jersey Governor Quits, Comes Out as Gay." *CNN*, August 13, 2004. http://edition.cnn .com/2004/ALLPOLITICS/08/12/mcgreevey.nj/.

"NumbersUSA Activists Squash Amnesty in Senate: Senate Rejects Cloture on Amnesty Bill 46–53." *PR Newswire*, June 28, 2007.

"Obama Assures Modi on Concerns Over H-1B Visa Issue." *Times of India*, January 26, 2015. http://timesofindia.indiatimes.com/india/Obama-assures-Modi-on-concerns-over-H-1B -visa-issue/articleshow/46022377.cms.

Obama, Barack. "Remarks of President Barack Obama: Address to Joint Session of Congress." *Whitehouse.gov*, February 24, 2009. www.whitehouse.gov/the-press-office/remarks-president -barack-obama-address-joint-session-congress.

——. "Statement of Administration Policy: H.R. 6429—STEM Jobs Act of 2012." *American Presidency Project*, November 28, 2012. www.presidency.ucsb.edu/ws/?pid=102707.

"Obama Lowers Temperature Against Outsourcing." *Economic Times*, March 28, 2009.

O'Connor, Patrick. "Anti-immigration Groups Up Against Unusual Coalition." *The Hill*, February 28, 2006.

Office of the Press Secretary. "Fact Sheet: Immigration Accountability Executive Action." *White House*, November 20, 2014. https://obamawhitehouse.archives.gov/the-press-office /2014/11/20/fact-sheet-immigration-accountability-executive-action.

——. "Leveling the Playing Field: Curbing Tax Havens and Removing Tax Incentives for Shifting Jobs Overseas." *White House*. Accessed March 26, 2015. https://www.whitehouse.gov /node/2739.

Olson, Elizabeth. "Congress Raises Limit on Skilled-Work Visas." *International Herald Tribune*, November 24, 2004.

Olson, Mancur. *The Rise and Decline of Nations: Economic Growth, Stagflation, and Social Rigidities.* New Haven, CT: Yale University Press, 1982.

"One Year of Startup India: Report Card." *TechCircle*, January 16, 2017.

Organisation for Economic Co-operation and Development (OECD). *Education at a Glance 2014: OECD Indicators.* Paris: OECD, 2014.

——. *The Measurement of Scientific and Technological Activities (Oslo Manual).* Paris: OECD, 1997.

——. "Population." *OECD.Stat*, March 23, 2016. http://stats.oecd.org/Index.aspx?DatasetCode =POP_FIVE_HIST.

O'Riain, Sean. *The Politics of High Tech Growth: Developmental Network States in the Global Economy.* Cambridge: Cambridge University Press, 2004.

"ORISE Workforce Studies Infographics—StayRates." *Oak Ridge Institute for Science and Education*, March 31, 2017. https://public.tableau.com/views/ORISEWorkforceStudiesInfo graphics-StayRates-mobilefriendly/5-YearStayRates?%3Aembed=y&%3AshowVizHome =no&%3Adisplay_count=y&%3Adisplay_static_image=y&%3AbootstrapWhenNotified=true.

Orleans, Leo A. "China's Changing Attitude Toward the Brain Drain and Policy Toward Returning Students." *China Exchange News* 17, no. 2 (1989): 2–5.

Papademetriou, Demetrios G., and Stephen Yale-Loehr. "Balancing Interests: Rethinking U.S. Selection of Skilled Immigrants." Washington, DC: Carnegie Endowment for International Peace, 1996.

Park, Haeyoun. "How Outsourcing Companies Are Gaming the Visa System." *New York Times*, November 10, 2015.

Paul, Joseph G., and Frank Caruso. "One of Trump's Biggest Plans to Stimulate the Economy Won't Be Great for Most Americans." *Business Insider*, June 1, 2017. www.businessinsider.com. au/alliancebernstein-on-trump-tax-plan-2017-5?r=US&IR=T.

Paulson, Amanda, Faye Bowers, and Daniel Wood. "To Immigrants, US Reform Bill Is Unrealistic." *Christian Science Monitor*, May 21, 2007.

Pavitt, Keith. "Innovation Processes." In *The Oxford Handbook of Innovation*, edited by Jan Fagerberg, David Mowery, and Richard R. Nelson. Oxford: Oxford University Press, 2006.

Pear, Robert. "Clinton Asks Congress to Raise the Limit on Visas for Skilled Workers." *New York Times*, May 12, 2000.

——. "Little-Known Group Claims a Win on Immigration." *New York Times*, July 15, 2007.

——. "U.S. High-Tech Firms Stymied on Immigration for Skilled Workers." *New York Times*, June 25, 2007. www.nytimes.com/2007/06/25/technology/25iht-visas.4.6326165.html.

Peters, Margaret E. "Trade, Foreign Direct Investment, and Immigration Policy Making in the United States." *International Organization* 68, no. 4 (2014): 811–44.

Peterson, Brenton D., Sonal S. Pandya, and David Leblang. "Doctors with Borders: Occupational Licensing as an Implicit Barrier to High Skill Migration." *Public Choice* 160, no. 1–2 (July 2014): 45–63.

Phillips, Kate. "Business Lobbyists Call for Action on Immigration." *New York Times*, April 15, 2006. www.nytimes.com/2006/04/15/us/15lobby.html.

Pietrucha, Bill. "Labor Challenges High Tech Job Shortage Claims." *Newsbytes News Network*, March 19, 1998.

"PM for Reverse Brain Drain of Scientists." *Economic Times*, January 4, 2011. http://articles.economictimes.indiatimes.com/2011-01-04/news/28424740_1_scientists-of-indian-origin-talent-pool-98th-indian-science.

Posner, Michael. "Groups Jockey for Position on Possible Boost in H-1B Visas." *Congress Daily*, November 17, 2004.

Potter, Mark, and Rich Philips. "Six Months after Sept. 11, Hijackers' Visa Approval Letters Received." *CNN*, March 13, 2002. http://edition.cnn.com/2002/US/03/12/inv.flight.school.visas/.

Powell, Colin. "Remarks at the Elliott School of International Affairs." Speech at George Washington University, Washington, DC, September 5, 2003. https://2001-2009.state.gov/secretary/former/powell/remarks/2003/23836.htm.

"Qianren Jihua [Thousand Talents Plan]." *Qianren Jihua Wang*, 2016. www.1000plan.org/qrjh/section/2.

Qin, Fei. "Global Talent, Local Careers: Circular Migration of Top Indian Engineers and Professionals." *Research Policy* 44, no. 2 (2015): 405–20.

Qualcomm. *Qualcomm Annual Report 2016*. San Diego, CA: Qualcomm, 2016. http://investor.qualcomm.com/annuals-proxies.cfm.

Quan, Xiaohong. "Knowledge Diffusion from MNC R&D Labs in Developing Countries: Evidence from Interaction Between MNC R&D Labs and Local Universities in Beijing." *International Journal of Technology Management* 51, no. 2 (2010): 364–86.

Rao, Nirupama. "America Needs More High-Skilled Worker Visas." *USA Today*, April 14, 2013. www.usatoday.com/story/opinion/2013/04/14/india-trade-technology-column/2075159/.

Ravenhill, John. "The Economics-Security Nexus in the Asia-Pacific region." In *Security Politics in the Asia-Pacific: A Regional-Global Nexus?*, edited by William Tow, 188-207. New York: Cambridge University Press, 2009.

——. "US Economic Relations with East Asia: From Hegemony to Complex Interdependence?" In *Bush and Asia: America's Evolving Relations with East Asia*, edited by Mark Beeson, 42-63. London: Routledge, 2006.

"RBI Eases Norms for Foreign Investment in Startups." *TechCircle*, October 21, 2016.

"R&D Technology Center China." *GE Lighting Asia Pacific*. Accessed April 16, 2014. www.gereveal.ca/LightingWeb/apac/resources/world-of-ge-lighting/research-and-development/china-technology-centre.jsp.

Reinsch, William. "What Is to Be Done on Trade?" *Stimson Spotlight*, June 7, 2016. www.stimson.org/content/what-be-done-trade.

"Rep. Tancredo Slams Senate's Compromise on Amnesty." *U.S. Fed News*, May 11, 2006.

"Research and Development." *General Electric*, 2017. www.ge.com/in/about-us/research-and-development.

Reuveny, Rafael, and William R. Thompson. *Growth, Trade, and Systemic Leadership*. Ann Arbor: University of Michigan Press, 2009.

Robbins, Liz. "New U.S. Rule Extends Stay for Some Foreign Graduates." *New York Times*, March 9, 2016. www.nytimes.com/2016/03/09/nyregion/new-us-rule-extends-stay-for-some-foreign-graduates.html.

Rodriguez, Cindy. "Congress Drops Plan to Bar Foreign Students." *Knight-Ridder Tribune Business News*, November 23, 2001.

——. "Foreign Workers Bill Approved." *Boston Globe*, October 4, 2000.

——. "Proposed Visa Ban Dropped." *Boston Globe*, November 23, 2001.

Romer, Paul M. "Endogenous Technological Change." *Journal of Political Economy* 98, no. 5 (1990): S71–S102.

Romm, Tony. "Apple Takes Washington." *Politico*, August 27, 2015. http://politi.co/1Px6AW0.

——. "Finally, Silicon Valley and Donald Trump Agree on Something: Taxes." *Recode*, April 26, 2017. www.recode.net/2017/4/26/15437330/silicon-valley-tech-donald-trump-agree-tax-repatriation-reform.

——. "How Silicon Valley Is Trying to Topple Trump—Beginning with a Special Election in Montana." *Recode*, May 25, 2017. www.recode.net/2017/5/25/15686802/silicon-valley-trump-montana-tech-for-campaigns.

Rosario, Katherine. "Five Simple Signs the Senate Immigration Bill Is Bad News," April 17, 2013. http://heritageaction.com/2013/04/5-simple-signs-the-senate-immigration-bill-is-bad-news/.

Rosenbaum, Jessica F. "Exploiting Dreams: H-1B Visa Fraud, Its Effects, and Potential Solutions." *University of Pennsylvania Journal of Business Law* 13, no. 3 (2010): 797–816.

Rosenfeld, Steven. "The GOP's Vicious Internal War: Republican Establishment Trying to Exile Tea Partiers and Extremists." *AlterNet*, February 12, 2014. www.alternet.org/tea-party-and-right/gops-vicious-internal-war-republican-establishment-trying-exile-tea-partiers-and.

Rothenberg, Stuart. "Heeee's Back: The Fall and Rise of Sen. Trent Lott." *Roll Call*, May 22, 2006.

Roy, Paul. "Impact of U.S. Senate Bill on Outsourcing." *Mondaq Business Briefing*, July 31, 2013.

Rubinstein, Ellis. "China's Leader Commits to Global Science and Scientific Exchange." *Science*, October 6, 2000. www.sciencemag.org/careers/2000/10/chinas-leader-commits-global-science-and-scientific-exchange.

Ruhs, Martin. *The Price of Rights: Regulating International Labor Migration*. Princeton, NJ: Princeton University Press, 2013.

Ruiz, Neil G., Jill H. Wilson, and Shyamali Choudhury. "The Search for Skills: Demand for H-1B Immigrant Workers in U.S. Metropolitan Areas." Washington, DC: Brookings Institution, 2012.

Rulon, Richard. "Competing for Foreign Talent." *Legal Intelligencer*, December 15, 2004.

"Ruxuan Zhongguo Qianren Jihua Waizhuan Xiangmu de Zhuanjia Da 381 Ming [Three Hundred Eighty-One Individuals Selected for China's Thousand Talents Foreign Experts Program]." *Kexue Wang [Science Net]*, April 15, 2017. http://news.sciencenet.cn/htmlnews/2017/4/373557.shtm.

Sabato, Larry. *PAC Power: Inside the World of Political Action Committees.* New York: Norton, 1984.

———. *Paying for Elections: The Campaign Finance Thicket.* New York: Priority, 1989.

Saxenian, AnnaLee. *The New Argonauts: Regional Advantage in a Global Economy.* Cambridge, MA: Harvard University Press, 2006.

Sazabo, Joan. "Opening Doors for Immigrants." *Nation's Business,* August 1, 1989.

Schemo, Diana Jean. "Problems Slow Tracking of Students from Abroad." *New York Times,* March 23, 2003.

Scheve, Kenneth F., and Matthew Jon Slaughter. *Globalization and the Perceptions of American Workers.* Washington, DC: Peterson Institute, 2001.

Schilling, Melissa. "Technology Shocks, Technological Collaboration, and Innovation Outcomes." *Organization Science* 26, no. 3 (May–June 2015): 668–86.

———. "Understanding the Alliance Data." *Strategic Management Journal* 30, no. 3 (2009): 233–60.

Schneider, Greg. "Anxious About Outsourcing; States Try to Stop U.S. Firms from Sending High-Tech Work Overseas." *Washington Post,* January 31, 2004.

Schouten, Fredreka. "Tech Firms Would Skirt Hiring Restrictions Under Deal." *USA Today,* May 21, 2013.

Schreiner, Bruce. "National Group Takes Aim at McConnell on Immigration." *Associated Press,* June 27, 2007.

Schroeder, Michael. "Business Coalition Battles Outsourcing Backlash." *Wall Street Journal,* March 1, 2004. www.wsj.com/articles/SB107809268846542227.

———. "States' Efforts to Curb Outsourcing Stymied: Business Groups Take the Lead in Weakening Attempts to Limit Work from Moving Abroad." *Wall Street Journal,* April 16, 2004.

———. "States Fight Exodus of Jobs: Lawmakers, Unions Seek to Block Outsourcing Overseas." *Wall Street Journal,* June 3, 2003.

Schroeder, Peter. "Extender Efforts Hit Roadblock as Senate Tables Tax Package." *Bond Buyer,* June 28, 2010.

Schröter, Harm G., and Anthony S. Travis. "An Issue of Different Mentalities: National Approaches to the Development of the Chemical Industry in Britain and Germany Before 1914." In *The Chemical Industry in Europe, 1850–1914,* edited by Ernst Homburg, Anthony S. Travis, and Harm G. Schröter, 95–120. Dordrecht, Netherlands: Kluwer, 1998.

Schumpeter, Joseph A. *Business Cycles: A Theoretical, Historical, and Statistical Analysis of the Capitalist Process.* New York: McGraw-Hill, 1939.

Schurenberg, Eric. "Why the Next Steve Jobs Could Be an Indian." *Mint,* October 28, 2011.

Schwaag-Serger, Sylvia. "Foreign Corporate R&D in China: Trends and Policy Issues." In *The New Asian Innovation Dynamics: China and India in Perspective,* edited by Govindan Parayil and Anthony P. D'Costa, 50–78. New York: Palgrave MacMillan, 2009.

Sell, Susan K. *Private Power, Public Law: The Globalization of Intellectual Property Rights.* Cambridge: Cambridge University Press, 2003.

Sen, Amiti. "India to Ask US for More H-1B Visas." *Economic Times,* October 19, 2009.

"Sen. Chuck Grassley to Place Hold on Employment-Based Visa Bill." *NumbersUSA,* November 30, 2011. www.numbersusa.com/content/news/november-30-2011/sen-chuck-grassley-place-hold-employment-based-visa-bill.html.

"Senator Feinstein Urges Major Changes in U.S. Student Visa Program." *Advocacy and Public Policymaking*, September 27, 2001. http://lobby.la.psu.edu/_107th/119_Student_Visas_Security /Congressional_Statements/Senate/S_Feinstein_09272001.htm.

"Senator Kerry Delivers Democratic Hispanic Radio Address." *U.S. Fed News*, April 1, 2006.

Service Contract Requirements for the Performance of Service Contracts Within the United States. Pub. L. No. 2005, c. 92 (New Jersey, 2005). www.njleg.state.nj.us/bills/BillView.asp.

Shachar, Ayelet. "Talent Matters: Immigration Policy-Setting as a Competitive Scramble Among Jurisdictions." In *Wanted and Welcome? Policies for Highly Skilled Immigrants in Comparative Perspective*, edited by Triadafilos Triadafilopoulos, 85–104. New York: Springer, 2013.

Shackelford, Brandon, and John Jankowski. "Information and Communications Technology Industries Account for $133 Billion of Business R&D Performance in the United States in 2013." *National Center for Science and Engineering Statistics*, April 2016. www.nsf.gov/statistics /2016/nsf16309/nsf16309.pdf.

Sharma, Dinesh C. *The Long Revolution: The Birth and Growth of India's IT Industry*. Noida: Harper Collins, 2009.

Sharma, Shumita. "US Offshore Outsourcing Ban Sparks Fears of Similar Laws." *Dow Jones International Newswires*, January 30, 2004.

Shear, Michael D., and Ashley Parker. "Boehner Is Said to Back Change on Immigration." *New York Times*, January 1, 2014. www.nytimes.com/2014/01/02/us/politics/boehner-is-said -to-back-change-on-immigration.html.

Sherman, Mark. "Feinstein Says Moratorium on Student Visas May Not Be Necessary." *Associated Press*, October 6, 2001.

Shesgreen, Deirdre. "Immigration Reform Critics Blast Boehner's Remarks." *Gannett News Service*, April 25, 2014.

Shi, Heping. "Beijing's China Card." *Harper's Magazine*, September 1990.

Shiver, Jube. "Alliance Fights Boost in Visas for Tech Workers." *Los Angeles Times*, August 5, 2000. http://articles.latimes.com/2000/aug/05/business/fi-64994.

Simmons, Joel W. *The Politics of Technological Progress*. Cambridge: Cambridge University Press, 2016.

Simon, Denis Fred, and Cong Cao. *China's Emerging Technological Edge: Assessing the Role of High-End Talent*. New York: Cambridge University Press, 2009.

Simons, John. "Impasse on Bill to Boost Visas Persists Between Firms, U.S." *Wall Street Journal*, August 6, 1998.

Simpson, Lori. *Engineering Aspects of Offshore Outsourcing*. Alexandria, VA: National Society of Professional Engineers, August 6, 2004. www.wise-intern.org/journal/2004/wise2004 -lorisimpsonfinalpaper.pdf.

Singh, J., and V. V. Krishna. "Trends in Brain Drain, Gain and Circulation: Indian Experience of Knowledge Workers." *Science Technology & Society* 20, no. 3 (November 1, 2015): 300–21.

Smith, Mark A. *American Business and Political Power: Public Opinion, Elections, and Democracy*. Chicago: University of Chicago Press, 2000.

——. "The Mobilization and Influence of Business Interests." In *The Oxford Handbook of American Political Parties and Interest Groups*, edited by L. Sandy Maisel and Jeffrey M. Berry, 451–67. Oxford: Oxford University Press, 2010.

Solon, Olivia. "US Tech Firms Bypassing Pentagon to Protect Deals with China, Strategist Says." *The Guardian*, March 2, 2016. www.theguardian.com/technology/2016/mar/02/us-tech-firms-pentagon-national-security-china-deals.

Southern Poverty Law Center. "John Tanton Is the Mastermind Behind the Organized Anti-immigration Movement." *Intelligence Report*, no. 106 (Summer 2002). www.splcenter.org/get-informed/intelligence-report/browse-all-issues/2002/summer/the-puppeteer?page=0,3.

Stanton, John, and Jennifer Yachnin. "Reid Plots to Block Conservatives." *Roll Call*, June 18, 2007.

Startz, Dick. "Sealing the Border Could Block One of America's Crucial Exports: Education." *The Brookings Institution* (January 31, 2017). www.brookings.edu/blog/brown-center-chalkboard/2017/01/31/sealing-the-border-could-block-one-of-americas-crucial-exports-education/.

State Council of the People's Republic of China. "Guojia Zhongchangqi Kexue He Jishu Fazhan Guihua Gangyao (2006–2020 Nian) [National Medium- and Long-Term Program for Science and Technology Development (2006–2020)]," *Zhongguo Zhengfu Menhu Wangzhan [Chinese Government Gateway Website]*, February 9, 2006. www.gov.cn/jrzg/2006-02/09/content_183787.htm.

——. "Guomin Jingji He Shehui Fazhan Dishierge Wunian Guihua Gangyao (Quan Wen) [Compendium of the Twelfth Five-Year Plan for Development of the National Economy and Society (Full Text)]," *Zhongguo Zhengfu Menhu Wangzhan [Chinese Government Gateway Website]*, March 16, 2011. www.gov.cn/2011lh/content_1825838.htm.

——. "Guowuyuan Guanyu Jiakuai Peiyu He Fazhan Zhanluexing Xinxing Chanye de Jueding [State Council Decision on Accelerating the Cultivation and Development of Strategic Emerging Industries]," *Zhongguo Zhengfu Menhu Wangzhan [Chinese Government Gateway Website]*, October 10, 2010. www.gov.cn/zwgk/2010-10/18/content_1724848.htm.

Steakley, Lia, Debra K. Rubin, and Peter Reina. "After 9/11, Overseas Students Find Foreigners Need Not Apply: Visa Application Hurdles Start to Ease but Long-Term Impacts Loom." *Engineering News-Record*, December 6, 2004.

Steinfeld, Edward S. *Playing Our Game: Why China's Rise Doesn't Threaten the West.* New York: Oxford University Press, 2010.

Steinhauer, Jennifer, Jonathan Martin, and David M. Herszenhorn. "Paul Ryan Calls Donald Trump's Attack on Judge 'Racist,' but Still Backs Him." *New York Times*, June 7, 2016. www.nytimes.com/2016/06/08/us/politics/paul-ryan-donald-trump-gonzalo-curiel.html.

Stephens, Paul. "International Students: Separate but Profitable." *Washington Monthly*, October 2013. www.washingtonmonthly.com/magazine/september_october_2013/features/international_students_separato46454.php?page=all.

Stokes, Bruce. "India's Paradox." *National Journal*, April 7, 2007.

Stuen, Eric T., Ahmed Mushfiq Mobarak, and Keith E. Maskus. "Skilled Immigration and Innovation: Evidence from Enrolment Fluctuations in US Doctoral Programmes." *The Economic Journal* 122, no. 565 (2012): 1143–76.

Swarns, Rachel L. "Senate, in Bipartisan Act, Passes an Immigration Bill." *New York Times*, May 26, 2006. www.nytimes.com/2006/05/26/washington/26immig.html.

Taylor, Mark Zachary. *The Politics of Innovation: Why Some Countries Are Better Than Others at Science and Technology.* Oxford: Oxford University Press, 2016.

Tea Party Patriots. "Senate Must Admit Full Costs of Immigration Bill Before Passing Another 'Train Wreck.'" *Tea Party Patriots,* May 6, 2013. www.teapartypatriots.org/all-issues/news /senate-must-admit-full-costs-of-immigration-bill-before-passing-another-train-wreck/.

"Technology Leaders Urge U.S. Senate to Approve Comprehensive Immigration Reform Legislation." *Information Technology Industry Council,* June 20, 2013. www.itic.org/news -events/news-releases/technology-leaders-urge-u-s-senate-to-approve-comprehensive -immigration-reform-legislation.

Teitelbaum, Michael S. *Falling Behind? Boom, Bust, and the Global Race for Scientific Talent.* Princeton, NJ: Princeton University Press, 2014.

Tellis, Ashley J., Janice Bially, Christopher Layne, Melissa McPherson, and Jerry M. Sollinger. *Measuring National Power in the Post-Industrial Age.* Santa Monica, CA: RAND, 2000.

Thibodeau, Patrick. "Ohio Bans Offshoring as It Gives Tax Relief to Outsourcing Firm." *Computerworld,* September 7, 2010. www.computerworld.com/article/2515465/it-outsourcing /ohio-bans-offshoring-as-it-gives-tax-relief-to-outsourcing-firm.html.

Thompson, Nicholas. "Obama vs. McCain: The Wired.com Scorecard." *Wired,* October 12, 2008. www.wired.com/2008/10/obama-v-mccain/.

Thompson, William R. "Long Waves, Technological Innovation, and Relative Decline." *International Organization* 44, no. 2 (1990): 201–33.

——. "Systemic Leadership, Evolutionary Processes, and International Relations Theory: The Unipolarity Question." *International Studies Review* 8, no. 1 (2006): 1–22.

Thomson Reuters. SDC Platinum Database. 2015. Access via subscription only.

Thoppil, Dhanya Ann, and Sean McLain. "Q&A: 'Parts of U.S. Visa Bill Discriminatory.'" *Wall Street Journal,* April 26, 2013.

Tichenor, Daniel J. *Dividing Lines: The Politics of Immigration Control in America.* Princeton, NJ: Princeton University Press, 2002.

Times Higher Education. "World University Rankings 2014–2015." *Times Higher Education,* 2014. www.timeshighereducation.co.uk/world-university-rankings/2014-15/world-ranking.

Tingley, Dustin H. "The Dark Side of the Future: An Experimental Test of Commitment Problems in Bargaining." *International Studies Quarterly* 55, no. 2 (June 2011): 521–44.

Trottman, Melanie. "Web Tool Could Help Boost Union Voter Turnout." *Wall Street Journal Online,* October 7, 2010.

Trumka, Richard. "Statement by AFL-CIO President Richard Trumka on Creating American Jobs and Ending Offshoring Act." *AFL-CIO,* September 28, 2010. www.aflcio.org/Press -Room/Press-Releases/Statement-by-AFL-CIO-President-Richard-Trumka-on-C8.

——. "Statement by AFL-CIO President Richard Trumka on the Promoting American Jobs and Closing Tax Loopholes Act." *AFL-CIO,* May 24, 2010. http://ftp.workingamerica.org/ Press-Room/Press-Releases/Statement-by-AFL-CIO-President-Richard-Trumka-on-t14.

Trumka, Richard, and Thomas J. Donohue. "Joint Statement of Shared Principles by U.S. Chamber of Commerce President and CEO Thomas J. Donohue and AFL-CIO President

Richard Trumka." *AFL-CIO*, February 21, *2013.* www.aflcio.org/Press-Room/Press-Releases /Joint-Statement-of-Shared-Principles-by-U.S.-Chamber-of-Commerce-President-and -CEO-Thomas-J.-Donohue-AFL-CIO-President-Richard-Trumka.

Trump, Donald J. "Presidential Executive Order on Buy American and Hire American." *White House,* April 18, 2017. www.whitehouse.gov/the-press-office/2017/04/18/presidential-executive -order-buy-american-and-hire-american.

Tucker, Bill. "Job Creation Stalls; Interview with Commerce Secretary Don Evans." *Lou Dobbs Tonight.* CNN, January 9, 2004.

Tumulty, Brian. "Small Manufacturers Aim Buy American Challenge at U.S. Job Losses." *Gannett News Service*, October 31, 2003. http://global.factiva.com/redir/default.aspx?P=sa& an=GNS0000020040107dzav000j7&cat=a&ep=ASE.

United Nations Educational, Scientific and Cultural Organization (UNESCO). "Education Data." *UIS. Stat*, 2014. www.uis.unesco.org/datacentre/pages/default.aspx.

——. "Global Flow of Tertiary-Level Students." *UIS. Stat*. Accessed December 10, 2014. www .uis.unesco.org/EDUCATION/Pages/international-student-flow-viz.aspx.

U.S. Citizenship and Immigration Services. *Characteristics of H-1B Specialty Occupation Workers: Fiscal Year 2015*. Washington, DC: U.S. Department of Homeland Security, 2016.

U.S. Department of Labor. "Fact Sheet 62: What Are 'Exempt' H-1B Nonimmigrants?" *Wage and Hour Division*, July 2008. www.dol.gov/whd/regs/compliance/FactSheet62/whdfs62Q.pdf.

U.S. Department of State. "Nonimmigrant Visa Issuances by Visa Class and by Nationality." *Travel. State. Gov*, 2016. http://travel.state.gov/content/visas/english/law-and-policy/statistics /non-immigrant-visas.html.

——. "Nonimmigrant Worldwide Issuance and Refusal Data by Visa Category." *Travel. State .Gov*, January 14, 2014. http://travel.state.gov/content/visas/english/law-and-policy/statistics /non-immigrant-visas.html.

U.S. General Accounting Office. *Assessment of the Department of Commerce's Report on Workforce Demand and Supply*. Washington, DC: U.S. General Accounting Office, 1998. http://gao.gov /assets/230/225415.pdf.

——. *Immigration and the Labor Market: Nonimmigrant Alien Workers in the United States*. Washington, DC: U.S. General Accounting Office, 1992. www.gao.gov/assets/160/151654.pdf.

——. *Improvements Needed to Reduce Time Taken to Adjudicate Visas for Science Students and Scholars*. Washington, DC: U.S. General Accounting Office, 2004. www.gao.gov/index .html.

U.S. Government Accountability Office. *Challenges in Attracting International Students to the United States and Implications for Global Competitiveness*. Washington, DC: U.S. Government Accountability Office, 2007. www.gao.gov/index.html.

——. *Performance of Foreign Student and Exchange Visitor Information System Continues to Improve, but Issues Remain*. Washington, DC: U.S. Government Accountability Office, 2005.

——. *Streamlined Visas Mantis Program Has Lowered Burden on Foreign Science Students and Scholars, but Further Refinements Needed.* Washington, DC: U.S. Government Accountability Office, 2005.

U.S. Immigration and Naturalization Service. *Report on Characteristics of Specialty Occupation Workers (H-1B): Fiscal Year 2000.* Washington, DC: U.S. Immigration and Naturalization Service, 2002.

"U.S. Led Effort Reaches 'Major Breakthrough' to Expand Information Technology Agreement." *Office of the United States Trade Representative,* July 2015. https://ustr.gov/about-us/policy-offices/press-office/press-releases/2015/july/us-led-effort-reaches-%E2%80%98major.

"US's Grassley—'Difficult' to Get Quick Action on Tax Bill." *Market News International,* June 22, 2004.

"Vajpayee Calls for Reversing Brain Drain, Cutting Red Tape." *Hindu Business Line,* January 4, 2003. www.thehindubusinessline.com/bline/2003/01/04/stories/2003010402410500.htm.

Valbrun, Marjorie, and Scott Thurm. "Foreign Workers Will Soon Get Fewer U.S. Visas." *Wall Street Journal,* October 1, 2003.

VandeHei, Jim, and Zachary A. Goldfarb. "Immigration Deal at Risk as House GOP Looks to Voters." *Washington Post,* May 28, 2006.

Vaughan, Martin. "Businesses Split Over Tax Credits." *Wall Street Journal,* August 4, 2010.

Vaughan, Martin, and Susan Davis. "Senate Ends Corporate Tax Debate for Now, OKs Outsourcing Deal." *Congress Daily,* March 5, 2004.

Vogel, David. *Fluctuating Fortunes: The Political Power of Business in America.* New York: Basic Books, 1989.

Wadhwa, Vivek. *The Immigrant Exodus: Why America Is Losing the Global Race to Capture Entrepreneurial Talent.* Philadelphia: Wharton Digital Press, 2012.

Wadhwa, Vivek, AnnaLee Saxenian, Richard B. Freeman, and Alex Salkever. *Losing the World's Best and Brightest: America's New Immigrant Entrepreneurs, Part V.* Kansas City, MO: Ewing Marion Kauffman Foundation, March 2009.

Wallace, Gregory, and Deirdre Walsh. "House Passes Immigration Bill to Keep Science and Technology Students in U.S." *CNN Wire,* November 30, 2012.

Walsten, Peter, Jia Lynn Yang, and Craig Timberg. "Facebook Flexes Political Muscle with Carve-Out in Immigration Bill." *Washington Post,* April 16, 2013.

Wang, Huiyao. *Rencai Zhanzheng [Talent War].* Beijing: China CITIC, 2009.

Wang, Huiyao, David Zweig, and Xiaohua Lin. "Returnee Entrepreneurs: Impact on China's Globalization Process." *Journal of Contemporary China* 20, no. 70 (June 2011): 413–31.

Wang, Jian, Lan Xue, and Zheng Liang. "Multinational R&D in China: From Home-Country-Based to Host-Country-Based." *Innovation: Management, Policy & Practice* 14, no. 2 (June 2012): 192–202.

Ward, David. "Letter to the Senate Judiciary Committee Regarding Feinstein Proposal on Student Visas." *American Association of Collegiate Registrars and Admissions Officers,* October 2, 2001. www.aacrao.org/advocacy/issues-advocacy/sevis.

Ward, David. Testimony. "Dealing with Foreign Students and Scholars in an Age of Terrorism: Visa Backlogs and Tracking Systems." *Hearing Before the U.S. House of Representatives Committee on Science.* 108th Cong., March 26, 2003.

——. "The Role of Technology in Preventing the Entry of Terrorists into the United States." *Hearing Before the Subcommittee on Technology, Terrorism, and Government Information of the Senate Judiciary Committee.* 107th Cong., October 12, 2001.

Wasem, Ruth Ellen. *Immigration: Legislative Issues on Nonimmigrant Professional Specialty (H-1B) Workers.* Washington, DC: Congressional Research Service, 2004.

——. *Immigration: Legislative Issues on Nonimmigrant Professional Specialty (H-1B) Workers.* Washington, DC: Congressional Research Service, 2007.

Washington Higher Education Secretariat. "About WHES." *Washington Higher Education Secretariat.* Accessed February 3, 2016. www.whes.org/index.html.

Wei, Yu, and Zhaojun Sun. "China: Building an Innovation Talent Program System and Facing Global Competition in a Knowledge Economy." *Academic Executive Brief,* 2012. http://academicexecutives.elsevier.com/articles/china-building-innovation-talent-program -system-and-facing-global-competition-knowledge.

Weisman, Jonathan. "Boehner Doubts Immigration Bill Will Pass in 2014." *New York Times,* February 6, 2014. www.nytimes.com/2014/02/07/us/politics/boehner-doubts-immigration -overhaul-will-pass-this-year.html.

——. "Bush, Adviser Assailed for Stance on 'Offshoring' Jobs." *Washington Post,* February 11, 2004.

——. "Immigration Bill Dies in Senate." *Washington Post,* June 29, 2007. www.washingtonpost .com/wp-dyn/content/article/2007/06/28/AR2007062800963.html.

Weisman, Jonathan, and Mark Kaufman. "Tax-Cut Bill Draws White House Doubts: Corporate Provisions Go Beyond 'Core Objective,' Treasury Secretary Says." *Washington Post,* October 5, 2004.

Weisman, Jonathan, and Jim VandeHei. "Immigration Bill Lobbying Focuses on House Leaders: With Senate in Hand, Bush May Face a Skeptical GOP Base." *Washington Post,* May 1, 2006.

Weiss, Linda. *America Inc.? Innovation and Enterprise in the National Security State.* Ithaca, NY: Cornell University Press, 2014.

Wong, Carolyn. *Lobbying for Inclusion: Rights Politics and the Making of Immigration Policy.* Stanford, CA: Stanford University Press, 2006.

Wong, Joseph. *Betting on Biotech: Innovation and the Limits of Asia's Developmental State.* New York: Cornell University Press, 2011.

World Bank Migration and Remittances Team. *Migration and Remittances: Recent Developments and Outlook.* Migration and Development Brief No. 23 (October 6, 2014). http://siteresources .worldbank.org/INTPROSPECTS/Resources/334934-1288990760745/Migrationand DevelopmentBrief23.pdf.

Worthen, Ben. "Regulations: What to Worry About." *CIO,* June 15, 2004.

Wright, Chris F. "Why Do States Adopt Liberal Immigration Policies? The Policymaking Dynamics of Skilled Visa Reform in Australia." *Journal of Ethnic and Migration Studies* 41, no. 2 (January 28, 2015): 306–28.

Wright, John R. *Interest Groups and Congress: Lobbying, Contributions and Influence.* Boston: Allyn and Bacon, 1996.

Yale-Loehr, Stephen, Demetrios G. Papademetriou, and Betsy Cooper. *Secure Borders, Open Doors: Visa Procedures in the Post–September 11 Era.* Washington, DC: Migration Policy Institute, 2005.

Ye, Min. *Diasporas and Foreign Direct Investment in China and India.* New York: Cambridge University Press, 2014.

Zengerle, Jason. "Silicon Smoothies." *New Republic*, June 8, 1998.

Zhao, Minyuan. "Conducting R&D in Countries with Weak Intellectual Property Rights Protection." *Management Science* 52, no. 8 (August 2006): 1185–99.

Zolberg, Aristide. *A Nation by Design: Immigration Policy in the Fashioning of America*. Cambridge, MA: Harvard University Press, 2006.

Zuckerberg, Mark. "Mark Zuckerberg: Immigrants Are the Key to a Knowledge Economy." *Washington Post*, April 10, 2013. www.washingtonpost.com/opinions/mark-zuckerberg -immigrants-are-the-key-to-a-knowledge-economy/2013/04/10/aba05554-a20b-11e2-82bc -511538ae90a4_story.html.

Zweig, David. "Learning to Compete: China's Efforts to Encourage a 'Reverse Brain Drain.'" In *Competing for Global Talent*, edited by Christiane Kuptsch and Eng Fong Pang, 187–214. Geneva: International Labor Office, 2006.

Zweig, David, and Changgui Chen. *China's Brain Drain to the United States: Views of Overseas Chinese Students and Scholars in the 1990s*. Berkeley: Institute of East Asian Studies, University of California, 1995.

Zweig, David, Chung Siu Fung, and Donglin Han. "Redefining the Brain Drain: China's 'Diaspora Option.'" *Science, Technology & Society* 13, no. 1 (2008): 1–33.

Zweig, David, and Huiyao Wang. "Can China Bring Back the Best? The Communist Party Organizes China's Search for Talent." *The China Quarterly* 215 (2013): 590–615.

INDEX